Systems & Control: Foundations & Applications

Series Editor
Christopher I. Byrnes, Washington University

Associate Editors
S.-I. Amari, University of Tokyo
B.D.O. Anderson, Australian National University, Canberra
Karl Johan Åström, Lund Institute of Technology, Sweden
Jean-Pierre Aubin, CEREMADE, Paris
H.T. Banks, University of Southern California, Los Angeles
John S. Baras, University of Maryland, College Park
A. Bensoussan, INRIA, Paris
John Burns, Virginia Polytechnic Institute, Blacksburg
Han-Fu Chen, Beijing University
M.H.A. Davis, Imperial College of Science and Technology, London
Wendell Fleming, Brown University, Providence, Rhode Island
Michel Fliess, CNRS-ESE, Gif-sur-Yvette, France
Keith Glover, University of Cambridge, England
Diederich Hinrichsen, University of Bremen, Federal Republic of Germany
Alberto Isidori, University of Rome
B. Jakubczyk, Polish Academy of Sciences, Warsaw
Hidenori Kimura, University of Osaka, Japan
Arthur J. Krener, University of California, Davis
H. Kunita, Kyushu University, Japan
Alexandre Kurzhansky, IIASA, Laxenburg, Austria and Academy of Sciences, U.S.S.R.
Harold J. Kushner, Brown University, Providence, Rhode Island
Anders Lindquist, Royal Institute of Technology, Stockholm
Andrzej Manitius, George Mason University, Fairfax, Virginia
Clyde F. Martin, Texas Tech University, Lubbock, Texas
Sanjoy Mitter, Massachusetts Institute of Technology, Cambridge
Giorgio Picci, University of Padova, Italy
Boris Pshenichnyj, Glushkov Institute of Cybernetics, Kiev
H.J. Sussman, Rutgers University, New Brunswick, New Jersey
T.J. Tarn, Washington University, St. Louis, Missouri
V.M. Tikhomirov, Institute for Problems in Mechanics, Moscow
Pravin P. Varaiya, University of California, Berkeley
Jan C. Willems, University of Gröningen, Sweden
W.M. Wonham, University of Toronto

Tamer Başar
Pierre Bernhard

H^∞-Optimal Control and Related Minimax Design Problems
A Dynamic Game Approach

1991

Birkhäuser
Boston • Basel • Berlin

Tamer Başar
Coordinated Science Laboratory
University of Illinois
Urbana, Illinois 61801
USA

Pierre Bernhard
Inria
Unité de Recherche Sophia-Antipolis
Valbone Cedex
France

Printed on acid-free paper.

© Birkhäuser Boston, 1991
All rights reserved. No part of this publication may be reproduced, stored in a retrieval system, or transmitted, in any form or by any means, electronic, mechanical, photocopying, recording or otherwise, without prior permission of the copyright owner.

ISBN 0-8176-3554-8
ISBN 3-7643-3554-8

Camera-ready copy prepaired by the author using TeX.
Printed and bound by Quinn-Woodbine, Woodbine, New Jersey
Printed in U.S.A.

9 8 7 6 5 4 3 2 1

Preface

One of the major concentrated activities of the past decade in control theory has been the development of the so-called "H^∞-optimal control theory," which addresses the issue of worst-case controller design for linear plants subject to unknown additive disturbances, including problems of *disturbance attenuation, model matching,* and *tracking*. The mathematical symbol "H^∞" stands for the *Hardy space* of all complex-valued functions of a complex variable, which are analytic and bounded in the open right-half complex plane. For a linear (continuous-time, time-invariant) plant, the H^∞ norm of the transfer matrix is the maximum of its largest singular value over all frequencies.

Controller design problems where the H^∞ norm plays an important role were initially formulated by *George Zames* in the early 1980's, in the context of sensitivity reduction in linear plants, with the design problem posed as a mathematical optimization problem using an (H^∞) operator norm. Thus formulated originally in the frequency domain, the main tools used during the early phases of research on this class of problems have been operator and approximation theory, spectral factorization, and (Youla) parametrization, leading initially to rather complicated (high-dimensional) optimal or near-optimal (under the H^∞ norm) controllers. Follow-up work in the mid 1980's have shown, however, that the maximum *McMillan degree* of these controllers is in fact in the order of the *McMillan degree* of the overall system transfer matrix, and further work has shown that in a time-domain characterization of these controllers (generalized) Riccati equations of the type that arises in linear-quadratic differential games play a key role. These findings have prompted further accelerated research on the topic, for more direct time-domain (state-space) derivation of these results — a direction which has also led to more general formulations, including time-varying plants and finite design horizons.

Among different time-domain approaches to this class of worst-case design problems, the one that uses the framework of *dynamic (differen-*

tial) *game theory* seems to be the most natural. This is so because the original H^∞-optimal control problem (in its equivalent time-domain formulation) is in fact a *minimax* optimization problem, and hence a *zero-sum game*, where the controller can be viewed as the *minimizing player* and disturbance as the *maximizing player*. Using this framework, we develop in this book a complete theory that encompasses continuous-time as well as discrete-time systems, finite as well as infinite horizons, and several different measurement schemes, including closed-loop perfect state, delayed perfect state, sampled state, closed-loop imperfect state, delayed imperfect state and sampled imperfect state information patterns.

It is our belief that the theory is now at a stage where it can easily be incorporated into a second-level graduate course in a control curriculum, that would follow a basic course in linear control theory covering LQ and LQG designs. The framework adopted in this book makes such an ambitious plan possible, and indeed, both authors will be teaching such courses during the Spring 1991 semester. The first author will incorporate this material into an optimal control course at the *University of Illinois*, and the second author will be teaching such a course at ESSI * and ISIA ** in Sophia Antipolis, and at *Ecole Polytechnique in Paris*.

For the most part, the only prerequisite for this book is a basic knowledge of linear control theory, at the level of, say, the texts by *Kwakernaak and Sivan (1972)* or *Anderson and Moore (1990)*. No background in differential games, or game theory in general, is required, as the requisite concepts and results have been developed in the book at the appropriate level. The book is written in such a way that makes it possible to follow the theory for the continuous- and discrete-time systems independently, so that a reader who is interested only in discrete-time systems, for example, may skip Chapters 4 and 5 without any disruption in the development of the theory. On the other hand, for the benefit of the reader who is interested in both continuous- and discrete-time results, we have included statements in each relevant chapter, that place the comparisons in proper perspective. The general framework adopted in this book, and the methodologies developed, should also make it possible for the interested reader to extend

---

* *Ecole Superieure des Sciences de l'Informatique, University of Nice.*
** *Institut Superieur d'Informatique et d'Automatique, Ecole des Mines de Paris.*

these results to more general classes of systems, such as nonlinear plants and/or distributed-parameter systems. We, in fact, present in the book some preliminary results on the former class of problems.

This project was conceived about six months ago at an International Symposium on *Differential Games and Applications*, held in Helsinki, Finland, and the collaboration between the two authors was facilitated by a ten-day visit of the second author to Urbana in October, and by one of the most incredible technological advances of the past decade — *the electronic mail*. The first author would like to thank his colleagues at the *Decision and Control Laboratory* of the *University of Illinois* (UI), *Urbana-Champaign*, for their encouragement and support in the writing of this book. Garry Didinsky, a Ph.D. student at UI, read parts of the manuscript, and besides providing valuable input, also contributed to the development of some of the results. Parts of the manuscript have also benefited from the comments of the participants at a one-day short course on the topic, offered by the first author at the *Coordinated Science Laboratory* of the *University of Illinois*, on October 2, 1990. The second author wishes to acknowledge, beyond the support of INRIA and the steadfast encouragement of its President, his share of the enthusiasm generated at the Coordinated Science Laboratory whose support and warm welcome during his stay in Urbana has been deeply felt. Finally, our thanks go to Francie Bridges, at the *Decision and Control Laboratory*, for her excellent assistance in word processing, which has transformed the hand-written version into the final form seen here.

Tamer Başar   Pierre Bernhard
Urbana, Illinois, USA   Sophia Antipolis, Valbonne, France
January 2, 1991   January 2, 1991

To

Tangül et Annette

Contents

Preface   v

1 A General Introduction to Minimax (H^∞-Optimal) Designs   1
    1.1   A Brief History   1
    1.2   A Relationship Between H^∞-Optimal Control and LQ Zero-Sum Dynamic Games   4
    1.3   Discrete- and Continuous-time Models   7
    1.4   Organization of the Book   9
    1.5   Conventions, Notation and Terminology   10

2 Basic Elements of Static and Dynamic Games   13
    2.1   Static Zero-Sum Games   13
    2.2   Discrete-Time Dynamic Games   16
    2.3   Continuous-Time Dynamic Games   24

3 The Discrete-Time Minimax Design Problem With Perfect State Measurements   28
    3.1   Introduction   28
    3.2   The Soft-Constrained Linear-Quadratic Dynamic Game . .   29
        3.2.1   Open-loop information structure for both players. .   30
        3.2.2   Closed-loop perfect state information for both players.   32
        3.2.3   Closed-loop 1-step delay information for both players.   38
        3.2.4   An illustrative example   42

3.3	Solution to the Disturbance Attenuation Problem	44
	3.3.1	General closed-loop information	44
	3.3.2	Illustrative example (continued)	48
	3.3.3	A least favorable distribution for the disturbance	50
	3.3.4	Optimum controller under the 1-step delay information pattern	52
3.4	The Infinite-Horizon Case	54
3.5	More General Classes of Problems	65
	3.5.1	More general plants and cost functions	65
	3.5.2	Nonzero Initial State	68
3.6	Extensions to Nonlinear Systems under Nonquadratic Performance Indices	69
3.7	Summary of Main Results	71

4 Continuous-Time Systems With Perfect State Measurements   72

4.1	Introduction	72
4.2	The Soft-Constrained Differential Game	73
	4.2.1	Open-loop information structure for both players.	74
	4.2.2	Closed-loop perfect state information for both players.	77
	4.2.3	Sampled-data information for both players.	80
	4.2.4	Delayed state measurements	84
4.3	The Disturbance Attenuation Problem	86
	4.3.1	Closed-loop perfect state information	87
	4.3.2	Sampled state measurements	89
	4.3.3	An illustrative example	90
	4.3.4	Delayed state measurements	92
4.4	The Infinite-Horizon Case	93
	4.4.1	The soft-constrained differential game	93
	4.4.2	The disturbance attenuation problem	103
	4.4.3	Illustrative example (continued)	107

	4.5	More General Classes of Problems	108
		4.5.1	A more general cost structure	109
		4.5.2	Unknown nonzero initial state	110
	4.6	Nonlinear Systems and Nonquadratic Performance Indices	111
	4.7	Main Points of the Chapter	112
5	**The Continuous-Time Problem With Imperfect State Measurements**			**114**
	5.1	Formulation of the Problem	114
	5.2	A Certainty Equivalence Principle, and Its Application to the Basic Problem $\mathcal{P}_\gamma$	117
	5.3	Sampled-Data Measurements	132
	5.4	The Infinite-Horizon Case	138
	5.5	More General Classes of Problems	144
		5.5.1	Cross terms in the cost function	144
		5.5.2	Delayed measurements	146
		5.5.3	Nonlinear/nonquadratic problems	148
	5.6	Main Results of the Chapter	149
6	**The Discrete-Time Problem With Imperfect State Measurements**			**151**
	6.1	The Problem Considered	151
	6.2	A Certainty Equivalence Principle and Its Application to the Basic Problem $\mathcal{P}_\gamma$	154
	6.3	The Infinite-Horizon Case	167
	6.4	More General Classes of Problems	172
		6.4.1	Cross terms in the performance index	172
		6.4.2	Delayed measurements	174
		6.4.3	The "filtering" problem	176
		6.4.4	Nonlinear/nonquadratic problems	178
	6.5	Main Points of the Chapter	178

7 Performance Levels For Minimax Estimators — 180

- 7.1 Introduction — 180
- 7.2 A Static Minimax Estimation Problem — 182
- 7.3 Optimum Performance Levels — 187
  - 7.3.1 Discrete time — 187
  - 7.3.2 Continuous time — 190
- 7.4 Summary of Main Results — 192

8 Appendix A: Conjugate Points — 194

9 Appendix B: Danskin's Theorem — 208

10 References — 213

11 List of Corollaries, Definitions, Lemmas, Propositions, Remarks and Theorems — 221

Chapter 1

A General Introduction to Minimax (H^∞-Optimal) Designs

1.1 A Brief History

A fundamental problem of theoretical and practical interest, that lies at the heart of control theory, is the design of controllers that yield acceptable performance for **not** a single plant and under known inputs, but rather a **family** of plants under various types of inputs and disturbances. The importance of this problem has long been recognized, and over the years various scientific approaches have been developed and tested [35]. A common initial phase of all these approaches has been the formulation of a mathematically well-defined problem, usually in the form of optimization of a performance index, which is then followed by either the use of available tools or the development of requisite new mathematical tools for the solution of these problems. Two of these approaches, *sensitivity approach* and the *linear-quadratic-Gaussian* (LQG) *design*, have dominated the field in the 1960's and early 1970's, with the former allowing small perturbations around an adopted nominal model ([32], [42]), and the latter ascribing some statistical description (specifically, Gaussian statistics) to the disturbances or unknown inputs [3]. During this period, the role of *game theory* in the design of robust (minimax) controllers was also recognized ([36], [90], [91], [75]), with the terminology "minimax controller" adopted from the statistical decision theory of the 1950's [76]. Here the objective is to obtain a design that minimizes a given performance index under *worst* possible disturbances or parameter variations (which maximize the same performance index). Since the desired controller will have to have a *dynamic structure*, this game-theoretic approach naturally requires the setting of dynamic (differential) games, but differential game theory (particularly with regard to information structures) being in its infancy in the 1960's, these initial attempts have not led to sufficiently general constructive methods for the design of robust controllers.

Research on dynamic game theory and information structures has made important strides in the 1970's (see, for example, the text [17]), and together with the sustained developments of the 1980's, it has improved our understanding of the intricate plays between solution concepts and dynamic information patterns in zero-sum and nonzero-sum differential games. Applications of the time-domain worst-case design philosophy during those years have been mostly to estimation and stochastic control problems with unstructured uncertainty, using the framework and tools of statistical decision theory (as in [79], [16], [70], [68], [66]), or using dynamic programming ([27], [29]), as well as to problems with structured (parametrized) uncertainties (for example, [63], [84], [67]), and to problems with only measurement uncertainty [89]. The worst-case approach has also been used in the stabilization of partially unknown plants, with the main tool there being the construction of appropriate Lyapunov functions (see, for example, [58]), and in the study of max-min controllability of linear systems ([20], [22], [46], [47]). The 1970's have also witnessed a burst of interest in yet a *fourth* approach to controller design, that of *adaptive control*, in both deterministic ([57]) and stochastic ([54]) domains, where the general objective is to *learn* the values of the disturbances and the unknown parameters through appropriately designed feedback or feedforward loops, and to achieve *asymptotic* optimality in performance.

Yet a *fifth* approach to controller design, H^∞-optimization, has dominated control theory research in the 1980's. Initiated by the pioneering work [93] at the beginning of the decade, it can also be viewed as a worst-case design methodology, but this time (as originally formulated) in the frequency domain, which is applicable to the three basic classes of design problems, namely *tracking, disturbance rejection*, and *model matching*. Here the objective is to obtain a controller that minimizes the maximum norm (i.e., the H^∞ norm) of an input-output operator, where the maximum is taken over the unknowns, such as disturbances. The derivation of the actual minimizing (optimal) controller being rather intractable, interest was first focussed on the characterization of all controllers that achieve a certain H^∞ norm bound. The main tools during this initial phase of research have been operator and approximation theory, spectral factorization and (*Youla*) parametrization, leading to optimum designs primarily in the frequency domain ([41], [31], [39], [40], [73]). Initially, it was thought that the corresponding controllers required high dimensional representations, but later work has shown that the maximum *McMillan* degree of these controllers is in fact in the order of the *McMillan* degree of the overall system transfer matrix [61]; furthermore, it has been shown that in a time-domain characterization of these controllers a (generalized) Riccati equation of the type that arises in linear-quadratic zero-sum differential games or risk-sensitive linear-exponentiated quadratic stochastic control plays a key role ([37], [43], [53], [69], [62], [44]). These correspondences have prompted further accelerated interest on the topic, with more recent work devoted to making the relationships between H^∞-optimal control, linear-quadratic differential games, and risk sensitive (entropy minimizing) control more precise, and searching for more direct time-domain derivations for (and thereby better insight into) the H^∞-optimal control problems ([81], [59], [50]). Such a time-domain derivation was in fact first given in [65] using a game-theoretic approach (and using some earlier results from [18]), which provided the solution to the (infinite-horizon) continuous-time H^∞-optimal control problem with perfect state measurements — much before the problem was originally formulated in the frequency domain. A related time-domain based worst-case controller synthesis has also appeared in the early 1970's in [26]. It should be noted that, as compared with the frequency-domain formulation, the time-domain approach is more encompassing, since it allows one to formulate also finite-horizon and time-varying versions of the original problem, and thus also study the transient behavior.

Among several different time-domain approaches to this controller design problem, the one that utilizes the tools of dynamic (or differential) game theory seems to be the simplest and the most intuitive, since after all the original H^∞-optimal control problem is a minimax optimization problem, and hence a zero-sum game (as to be clarified shortly). Accordingly, the main purpose of this book is to provide a self-contained exposition to the essential elements of (and relevant results from) linear-quadratic (LQ) zero-sum dynamic game theory, and to show how these tools can directly be used in solving the H^∞-optimal control problem in both continuous and discrete time, for finite and infinite horizons, and under different perfect and imperfect state information patterns. The work presented here is based on partly independent and partly joint recent work of the authors on the topic. Some selected results have been reported earlier in [5], [6], [11], [9], [8], [14], [12], [10], [23], [25].

1.2 A Relationship Between H^∞-Optimal Control and LQ Zero-Sum Dynamic Games

For the basic worst-case controller design problem (in continuous or discrete time) let us adopt the following compact notation:

$$\left. \begin{array}{rcl} z & = & G_{11}(w) + G_{12}(u) \\ y & = & G_{21}(w) + G_{22}(u) \\ u & = & \mu(y) \end{array} \right\}. \tag{1.1}$$

Here z, y, w and u denote, respectively, the controlled output, measured output, disturbance, and control variables, and they belong to appropriate

Hilbert spaces, denoted $\mathcal{H}_z$, $\mathcal{H}_y$ $\mathcal{H}_w$ and $\mathcal{H}_u$, respectively. G_{ij}, $i,j = 1,2$, are appropriate bounded causal linear operators, and so is $\mu \in \mathcal{M}$, which is called the *controller*. Here $\mathcal{M}$ is the controller space, which is assumed to be compatible with the information available to the controller. For every fixed $\mu \in \mathcal{M}$ we can introduce a bounded causal linear operator $T_\mu : \mathcal{H}_w \to \mathcal{H}_z$, defined by

$$T_\mu(w) = G_{11}(w) + G_{12}(I - \mu \circ G_{22})^{-1}(\mu \circ G_{21})(w)$$

where $\mu \circ G_{2i}$ denotes a composite operator, and we assume here that the required inverse exists. (A sufficient condition for this would be for either μ or G_{22} to be strictly causal, which we henceforth assume.) Furthermore, implicit in the definition of the controller space $\mathcal{M}$ is the requirement that, for infinite-horizon problems, every $\mu \in \mathcal{M}$ internally stabilizes the underlying system. The design objective is to optimally attenuate the disturbance at the output, which, in mathematical terms, is the optimization problem,

$$\inf_{\mu \in \mathcal{M}} \ll T_\mu \gg =: \gamma^* \qquad (1.2a)$$

where $\ll \cdot \gg$ denotes the operator norm of T_μ, i.e.,

$$\ll T_\mu \gg := \sup_{\|w\|_w \leq 1} \|T_\mu(w)\|_z \equiv \sup_{w \in \mathcal{H}_w} \|T_\mu(w)\|_z / \|w\|_w \qquad (1.2b)[1]$$

Here $\|\cdot\|_{(.)}$ denotes the appropriate Hilbert space norm, with the subscript identifying the corresponding space. At places where there is no ambiguity from context, we will drop this identifying subscript. Now, in view of (1.2b), the optimization problem (1.2a) clearly defines a "min max" problem, and a natural question to ask is whether the "infimum" and "supremum" operators can be interchanged in the construction of these controllers. In the parlance of game theory, this question is translated into one of the equivalence of upper and lower values of a zero-sum game defined by the kernel

---

[1] The identity here is valid because μ is a linear controller. If μ is allowed to be a nonlinear controller, it then becomes more appropriate to work with the second expression.

$\|T_\mu(w)\|_z/\|w\|_w$, a topic which will be studied later. We should note at this point the obvious inequality (see Section 2.1 for details),

$$\overbrace{\inf_{\mu\in\mathcal{M}}\sup_{w\in\mathcal{H}_w}\|T_\mu(w)\|_z/\|w\|_w}^{upper\ value} \geq \overbrace{\sup_{w\in\mathcal{H}_w}\inf_{\mu\in\mathcal{M}}\|T_\mu(w)\|_z/\|w\|_w}^{lower\ value}, \quad (1.3)$$

and also point out to the fact that for a large class of problems to be studied in this book this inequality is in fact strict, unless the disturbance is allowed to be a stochastic process (see Section 3.3.3 for a demonstration of this point). Toward formulating a different (in many ways, a simpler) game associated with the worst-case design problem formulated above, let us first assume that there exists a controller $\mu^* \in \mathcal{M}$ satisfying the minimax disturbance attenuation bound γ^* in (1.2a). Then, (1.2a) becomes equivalent to :

(i)
$$\|T_{\mu^*}(w)\|_z^2 \leq {\gamma^*}^2 \|w\|_w^2, \text{ for all } w \in \mathcal{H}_w \quad (1.4a)$$

and

(ii) there is no other $\mu \in \mathcal{M}$ (say, $\hat{\mu}$), and a corresponding $\hat{\gamma} < \gamma^*$, such that
$$\|T_{\hat{\mu}}(w)\|_z^2 \leq \hat{\gamma}^2 \|w\|_w^2, \quad \text{for all } w \in \mathcal{H}_w. \quad (1.4b)$$

Now, introducing the parameterized (in $\gamma \geq 0$) family of cost functions:

$$J_\gamma(\mu, w) := \|T_\mu(w)\|_z^2 - \gamma^2\|w\|_w^2 \equiv J(\mu, w) - \gamma^2\|w\|_w^2 \quad (1.5)$$

(i) and (ii) above become equivalent to the problem of finding the smallest value of $\gamma \geq 0$ under which the upper value of the associated game with objective function $J_\gamma(\mu, w)$ is bounded, *and* finding the corresponding controller that achieves this upper value. It will turn out that for the class of systems under consideration here, the zero-sum dynamic game defined by the kernel (1.5) has equal upper and lower values (whenever they are finite), which makes the existing theory on saddle-point solutions of linear-quadratic dynamic games directly applicable to this class of worst-case design problems. The dynamic game whose performance index is given by $J_\gamma(\mu, w)$, as above, will be called the *soft-constrained game* associated with the disturbance attenuation problem. The terminology "soft-constrained" is used to reflect the fact that in this game there is no hard bound on the disturbance w, while in the original problem characterized by the kernel $J(\mu, w)$ (see (1.5)) a norm bound had to be imposed on w. In both cases, the underlying dynamic optimization problem is a *two-person zero-sum dynamic game*, with the controller (u) being the minimizing player (henceforth called Player 1), and the disturbance being the maximizing player (called Player 2).

1.3 Discrete- and Continuous-time Models

The models we will work with in this book are the following state-space representations of (1.1) in discrete and continuous time:

Discrete Time :

$$x_{k+1} = A_k x_k + B_k u_k + D_k w_k, \qquad (1.6a)$$

$$z_k = H_k \underbrace{(\hat{H}_k x_k + \hat{G}_{k-1} u_{k-1} + \hat{F}_{k-1} w_{k-1})}_{\zeta_k} + G_k u_k, \qquad (1.6b)[2]$$

$$y_k = C_k x_k + E_k w_k, \quad k = 1, 2, \ldots \qquad (1.6c)$$

$$L(u, w) = |\zeta_{K+1}|^2_{Q_f} + \sum_{k=1}^{K} |z_k|^2 \; ; \quad Q_f \geq 0, \qquad (1.7a)$$

$$L_\gamma(u, w) = L(u, w) - \gamma^2 \sum_{k=1}^{K} |w_k|^2 \qquad (1.7b)$$

---

[2] The reason why we write the "controlled output" z as in (1.6b) is to be able to consider different special cases in Chapters 3 and 6, without introducing additional notation.

where upper case letters denote matrices of appropriate dimensions, $\{x_k\}$ is the state vector sequence, $|\cdot|$ denotes an appropriate Euclidean (semi-) norm, and K is either finite or $+\infty$. Furthermore,

$$u_k = \mu_k(y_{[1,k]}), \quad y_{[1,k]} := \{y_1, \ldots, y_k\}, \tag{1.8}$$

where the *control law* (or, equivalently, *control policy* or *strategy*) $\mu := \mu_{[1,K]} \in \mathcal{M}$ is taken as some linear mapping, even though we will establish some results also for a class of nonlinear controllers. We will also consider the cases where some structural constraints are imposed on the dependence of the control on the measurement sequence $y_{[1,k]}$ (such as delayed measurements). For any admissible control law $\mu_{[1,K]}$, if the *control action* generated by (1.8) is substituted into (1.7b), we will denote the resulting objective function as $J_\gamma(\mu, w)$, which is defined over $\mathcal{M} \times \mathcal{H}_w$ (whereas L_γ was defined on $\mathcal{H}_u \times \mathcal{H}_w$).[3] This corresponds to (1.5) defined in the previous section. Likewise, the counterpart of L (given by (1.7a)) will be denoted by J. Note that the functional J is in fact J_γ evaluated at $\gamma = 0$; the same interpretation applies to the relationship between L and L_γ.

Continuous Time :

$$\dot{x} = A(t)x + B(t)u(t) + D(t)w(t), \quad x(0) = x_0 \tag{1.9a}$$

$$z(t) = H(t)x(t) + G(t)u(t) \tag{1.9b}$$

$$y(t) = C(t)x(t) + E(t)w(t) \tag{1.9c}$$

$$L(u,w) = |x(t_f)|^2_{Q_f} + \int_0^{t_f} |z(t)|^2 \, dt \;; \quad Q_f \geq 0\,, \tag{1.10a}$$

$$L_\gamma(u,w) = \widetilde{L}(u,w) - \gamma^2 \int_0^{t_f} |w(t)|^2 \, dt \tag{1.10b}$$

$$u(t) = \mu(t, y_{[0,t]}), \quad t \geq 0 \tag{1.11}$$

---

[3] Clearly, J_γ is identical with the functional L_γ when the control is restricted to be open-loop. Since J_γ is more general than L_γ, we will mostly be using the notation J_γ (and correspondingly J) in this book, even when working with action variables, unless the specific context requires the difference between the two to be explicitly recognized.

where $\mu \in \mathcal{M}$ is again a linear controller, compatible with the underlying information structure (as to whether continuous or sampled measurements are made, whether there is delayed dependence on $y_{[0,t]}$, etc.). Again, we will use the notation $J_\gamma(\mu, w)$, to denote the counterpart of the performance index (1.10b) (or equivalently (1.5)), defined over the product space $\mathcal{M} \times \mathcal{H}_w$, and also use $J(\mu, w)$ to denote the counterpart of (1.10a). We will study in this book both the finite horizon (finite t_f) and infinite horizon ($t_f \to \infty$) cases, and some of the results to be presented will be valid even in the general class of nonlinear controllers. In both discrete and continuous time, the initial states (x_1 and x_0) will either be taken to be zero, or be taken as part of the disturbance, and in the latter case an additional negative term will be added to (1.7b) and (1.10b), which will involve Euclidean norms of x_1 and x_0, respectively. We postpone the discussion on the introduction of these norms into the performance indices until Sections 3.5 and 4.5.

1.4 Organization of the Book

In the next chapter, we introduce some basic notions from zero-sum static and dynamic game theory, and present some key results on the existence, uniqueness, and characterization of saddle-point equilibria of general games. Chapter 3 deals with the class of discrete-time linear-quadratic dynamic games with soft constraints and under various perfect state information patterns, and obtains necessary and sufficient conditions for boundedness of the upper value. These results are then applied to the discrete-time H^∞-optimal control problem so as to obtain optimal controllers under perfect state and delayed state information patterns. Both finite- and infinite-horizon problems are addressed, and in the latter case the stabilizability of the optimal controller is established under appropriate conditions. Chapter 4 presents the counterparts of these results in the continuous time, and particularly deals with the perfect state and sampled state information patterns. Complete extensions to the imperfect information case are developed in the next two chapters. The first of these (Chapter 5) is devoted to the class of continuous-time linear-quadratic game and minimax design problems, under both continuous and sampled imperfect state measurements. Chapter 6, on the other hand, develops a complete set of results for the discrete-time problem with imperfect (disturbance corrupted) state measurements. Chapter 7 discusses a class of related minimax design problems in filtering and smoothing. This is followed by two appendices (Chapters 8 and 9) that present some useful results on conjugate points, which are extensively used in the developments of Chapters 4 and 5, and a technical result needed in the proof of Theorem 5.1 in Chapter 5. The book ends with a list of references, and a table that indicates the page numbers of the Lemmas, Theorems, Definitions, etc. appearing in the text.

1.5 Conventions, Notation and Terminology

The book comprises seven chapters and two appendices. Each chapter is divided into sections, and sections occasionally into subsections. *Section 3.2*, for example, refers to the second section of *Chapter 3*, while *Section 3.2.3* is the third subsection of *Section 3.2*.

Items like theorems, definitions, lemmas, etc., are identified within each chapter according to the "telephone numbering system"; thus, *Theorem 4.2* would be the second theorem of *Chapter 4*, and *equation (5.8)* would be the eighth equation of *Chapter 5*.

The following symbols are adopted in the book, unless stated otherwise in a specific context:

$\mathbb{R}^n$	n-dimensional Euclidean space
$\mathbb{N}$	the set of natural numbers
:=	defined by
=:	defines
$\forall$	for all
$\diamond$	end of proof, remark, lemma, etc.

x	state variable; x_k in discrete time and $x(t)$ in continuous time
x_k' (A')	transpose of the vector x_k (of matrix A)
$\|x_k\|$	a Euclidean norm of the state vector x at time k
$\|\|x\|\|$	a Hilbert space norm of the state trajectory x
u	control variable; u_k in discrete time and $u(t)$ in continuous time
$\mathcal{U}$	space where the control trajectory lies; occasionally denoted by $\mathcal{H}_u$, $\mathcal{H}$ standing for *Hilbert space*
w	disturbance variable; w_k in discrete time and $w(t)$ in continuous time
$\mathcal{W}$	space where the disturbance trajectory lies; occasionally denoted by $\mathcal{H}_w$
$u_{[t_1,t_2]}$	control restricted to the discrete or continuous time interval $[t_1, t_2]$; also the notation u^t is used to denote the sequence $u_{[1,t]}$ in discrete time and $u_{[0,t]}$ in continuous time
y	measurement variable; y_k in discrete time and $y(t)$ in continuous time
$\mathcal{Y}$	space where the measurement trajectory lies
$\mu \in \mathcal{M}$	control policy, belonging to a given policy space $\mathcal{M}$
$\nu \in \mathcal{N}$	policy for disturbance, belonging to policy space $\mathcal{N}$
$\ell^2([t_1, t_2], I\!R^n)$	the Hilbert space of square summable functions on the discrete interval $[t_1, t_2]$, taking values in $I\!R^n$
$L^2([t_1, t_2], I\!R^n)$	the Hilbert space of square integrable functions on $[t_1, t_2]$, taking values in $I\!R^n$
J	performance index (cost function) of the original disturbance attenuation problem
J_γ	performance index of the associated soft-constrained dynamic game, parameterized by $\gamma > 0$

ARE	algebraic Riccati equation
RDE	Riccati differential equation
FB	feedback
LQ	linear-quadratic
CLPS	closed-loop perfect state
SDPS	sampled-data perfect state
DPS	delayed perfect state (in the continuous time)
CLD	closed-loop delayed (in the discrete time)
$\text{Tr}[\Lambda]$	trace of the square matrix Λ
$\rho(\Lambda)$	spectral radius of the square matrix Λ
Ker E	kernel of the matrix (or operator) E

Throughout the book, we will use the words "optimal" and "minimax" interchangeably, when referring to a controller that minimizes the maximum norm. Similarly, the corresponding attenuation level will be referred to as *optimal* or *minimax*, interchangeably. Also, the words *surjective* and *onto* (and likewise, *injective* and *one-to-one*) will be used interchangeably, the former more extensively than the latter.

Chapter 2

Basic Elements of Static and Dynamic Games

Since our approach in this book is based on (dynamic) game theory, it will be useful to present at the outset some of the basic notions of zero-sum game theory, and some general results on the existence and characterization of saddle points. We first discuss, in the next section, static zero-sum games, that is games where the actions of the players are selected independently of each other; in this case we also say that the players' strategies are *constants*. We then discuss in Sections 2.2 and 2.3 some general properties of dynamic games (with possibly nonlinear dynamics), first in the discrete time and then in the continuous time, with the latter class of games also known as differential games. In both cases we also introduce the important notions of *representation of a strategy, strong time consistency,* and *noise insensitivity*.

2.1 Static Zero-Sum Games

Let $L = L(u, w)$ be a functional defined on a product vector space $\mathcal{U} \times \mathcal{W}$, to be minimized by $u \in U \subset \mathcal{U}$ and maximized by $w \in W \subset \mathcal{W}$, where U and W are the constraint sets. This defines a zero-sum game, with kernel L, in connection with which we can introduce two values:

Upper value:

$$\bar{L} := \inf_{u \in U} \sup_{w \in W} L(u, w) \tag{2.1a}$$

Lower value:

$$\underline{L} := \sup_{w \in W} \inf_{u \in U} L(u, w) \tag{2.1b}$$

with the obvious inequality

$$\bar{L} \geq \underline{L}. \tag{2.1c}$$

If we have an equality in (2.1c), the common value

$$L^* = \bar{L} = \underline{L} \tag{2.2}$$

is called the *value* of the zero-sum game, and furthermore if there exists a pair $(u^* \in U, w^* \in W)$ such that

$$L(u^*, w^*) = L^* \tag{2.3}$$

then the pair (u^*, w^*) is called a (pure-strategy) *saddle-point solution*. In this case we say that the game admits a *saddle point* (in pure strategies). Such a saddle-point solution will equivalently satisfy the so-called "pair of saddle-point inequalities":

$$L(u^*, w) \leq L(u^*, w^*) \leq L(u, w^*), \quad \forall u \in \mathcal{U}, \forall w \in \mathcal{W}. \tag{2.4}$$

Not every zero-sum game admits a saddle point, or even a value. Consider, for example, the zero-sum game:

$$L(u, w) = (u - w)^2; \quad U = W = [0, 2]$$

for which

$$\underline{L} = 0, \quad \bar{L} = 1,$$

and hence the game does not have a value (in pure strategies). If, however, we take for w the two-point probability mass function p_w^* defined by

$$w^* = \begin{cases} 0 & \text{w.p.} \frac{1}{2} \\ 2 & \text{w.p.} \frac{1}{2} \end{cases}$$

and choose

$$u^* = 1, \quad \text{with probability (w.p.) } 1,$$

then
$$\min_{u \in [0,2]} \left[(u-0)^2 \frac{1}{2} + (u-2)^2 \frac{1}{2} \right] = 1$$

which is attained uniquely at $u = 1$, and

$$\max_{w \in [0,2]} \{(1-w)^2\} = 1$$

attained at $w = 0$ and $w = 2$, which is the support set of p_w^*. Hence, in the extended space of "mixed policies", the game admits a "mixed" saddle point. To define such a saddle point in precise terms, let

$$F(P,Q) = \iint_{U \times W} L(u,w) P(du) Q(dw) \tag{2.5a}$$

where P (respectively, Q) is a probability measure on U (respectively, W). The counterpart of (2.1c) now is

$$\bar{F} := \inf_{P} \sup_{Q} F(P,Q) \geq \sup_{Q} \inf_{P} F(P,Q) =: \underline{F} \tag{2.5b}$$

where $\bar{F}$ (respectively, $\underline{F}$) is the *upper* (respectively, *lower*) *value* in *mixed strategies (policies)*. If

$$\bar{F} = \underline{F} =: F^*, \tag{2.6a}$$

then F^* is the *value* in *mixed strategies*. Furthermore, if there exist probability measures (P^*, Q^*) such that

$$F(P^*, Q^*) = F^*, \tag{2.6b}$$

then (P^*, Q^*) is a *mixed saddle-point solution*. Some of the standard existence results on pure and mixed saddle points are the following (see [17] for proofs):

Theorem 2.1. *If U and W are finite sets, the zero-sum game admits a saddle point in mixed policies.* ◊

Theorem 2.2. *Let U, W be compact, and L be continuous in the pair (u, w). Then, there exists a saddle point in mixed policies.* ◇

Theorem 2.3. *In addition to the hypotheses of Theorem 2.2 above, let U and W be convex, $L(u, w)$ be convex in $u \in U$ for every $w \in W$, and concave in $w \in W$ for every $u \in U$. Then there exists a saddle point in pure policies. If, furthermore, L is <u>strictly</u> convex-concave, the saddle point solution is unique.* ◇

If the saddle-point solution of a zero-sum game is not unique, then any ordered combination of these multiple saddle-point equilibria could be adopted as a solution, in view of the following simple, but very useful property [17].

Property 2.1: Ordered interchangeability. *For a zero-sum game, $\{L; U, W\}$, if $(\bar{u}, \bar{w}) \in U \times W$ and $(\tilde{u}, \tilde{w}) \in U \times W$ are two saddle-point pairs, then the pairs $(\bar{u}, \tilde{w})$ and $(\tilde{u}, \bar{w})$ also constitute saddle points.* ◇

This ordered interchangeability property of multiple saddle points holds not only for *pure-strategy* saddle points, but also for *mixed-strategy* ones, in which case the action variables u and w are replaced by the corresponding probability measures P and Q. It also holds in the context of dynamic games, with the action variables now replaced by the corresponding strategies of the players.

2.2 Discrete-Time Dynamic Games

In this section we introduce a general class of finite-horizon discrete-time zero-sum games, discuss in this context various information patterns, and present a sufficient condition for the existence of a saddle point when the

Static and Dynamic Games

information pattern is perfect state. We also introduce the notions of *representation of a strategy, strong time consistency*, and (asymptotic) *noise insensitivity*.

Consider a zero-sum dynamic game described by the state equation

$$x_{k+1} = f_k(x_k, u_k, w_k), \quad k \in [1, K] := \{1, \ldots, K\} \tag{2.7a}$$

$$y_k = h_k(x_k, w_k), \quad k \in [2, K] \tag{2.7b}$$

and with the finite-horizon cost given by

$$L(u, w) = \sum_{k=1}^{K} g_k(x_{k+1}, u_k, w_k, x_k) \tag{2.8}$$

which is to be minimized by Player 1 and maximized by Player 2, using the control vectors $u_{[1,K]}$ and $w_{[1,K]}$, respectively. The state x_k takes values in $I\!R^n$, the measurement vector y_k takes values in $I\!R^p$, and the control vector of Player i takes values in $I\!R^{m_i}$, $i = 1, 2$. It is assumed (at this point) that the initial state x_1 is known to both players.

The formulation above will not be complete, unless we specify the information structure of the problem, that is the nature of the dependence of the control variables on the state or the measurement vector. In this book, we will be interested primarily in four different types of information[1]:

(i) *Closed-loop perfect state* (CLPS): The control is allowed to depend on the current as well as the entire past values of the state, i.e.,

$$\begin{aligned} u_k &= \mu_k(x_{[1,k]}), \quad k \in [1, K] \\ x_{[1,k]} &:= (x_1, \ldots, x_k) \end{aligned} \tag{2.9a}$$

where $\mu_{[1,k]} := (\mu_1, \ldots, \mu_k)$ is known as a (control) policy. We let $\mathcal{M}_{\text{CLPS}}$ denote the set of all such (Borel measurable) policies.

(ii) *Closed-loop 1-step delay* (CLD): The control is not allowed to depend on the current value of the state, but it can depend on the entire past

---

[1] Here we describe these from the point of view of Player 1; a similar listing could be given for Player 2 as well.

values of the state, i.e.,

$$
\begin{aligned}
u_k &= \mu_k(x_{[1,k-1]}), \quad k \in [2, K], \\
&= \mu_1(x_1), \quad k = 1.
\end{aligned}
\tag{2.9b}
$$

We denote the corresponding policy space by $\mathcal{M}_{\text{CLD}}$.

(iii) *Open-loop* (OL): The control depends only on the initial state, i.e.,

$$u_k = \mu_k(x_1), \quad k \in \mathcal{K}. \tag{2.9c}$$

The corresponding policy space is denoted by $\mathcal{M}_{\text{OL}}$.

(iv) *Closed-loop imperfect state* (CLIS): The control is allowed to depend on the current as well as the entire past values of the measurement vector, i.e.,

$$
\begin{aligned}
u_k &= \mu_k(y_{[2,k]}, x_1), \quad k \in [2, K], \\
&= \mu_1(x_1), \quad k = 1.
\end{aligned}
\tag{2.9d}
$$

We let $\mathcal{M}_{\text{CLIS}}$ denote the set of all such (Borel measurable) policies. Several delayed versions of this CLIS information pattern will also be used in this book; but we postpone their discussion until Chapter 6.

Let $\mathcal{M}$ denote the policy space of Player 1 under any one of the information patterns introduced above. Similarly let $\mathcal{N}$ denote the policy space of Player 2. Further, introduce the function $J : \mathcal{M} \times \mathcal{N} \to \mathbb{R}$ by

$$
\begin{aligned}
J(\mu, \nu) &= L(u, w), \quad u_k = \mu_k(\cdot), \quad w_k = \nu_k(\cdot), \quad k \in [1, K]; \\
(\mu_{[1,K]}, \nu_{[1,K]}) &\in \mathcal{M} \times \mathcal{N},
\end{aligned}
\tag{2.10}
$$

where we have suppressed the dependence on the initial state x_1. The triple $\{J; \mathcal{M}, \mathcal{N}\}$ constitutes the **normal form** of the zero-sum dynamic game, in the context of which we can introduce the notion of a *saddle-point equilibrium*, as in (2.4).

Definition 2.1. *Given a zero-sum dynamic game* $\{J; \mathcal{M}, \mathcal{N}\}$ *in normal form, a pair of policies* $(\mu^*, \nu^*) \in \mathcal{M} \times \mathcal{N}$ *constitutes a saddle-point solution if, for all* $(\mu, \nu) \in \mathcal{M} \times \mathcal{N}$,

$$J(\mu^*, \nu) \leq J^* := J(\mu^*, \nu^*) \leq J(\mu, \nu^*) \tag{2.11a}$$

The quantity J^* *is the value of the dynamic game.* ◊

The value is defined even if a saddle-point solution does not exist, as

$$\bar{J} := \inf_{\mu \in \mathcal{M}} \sup_{\nu \in \mathcal{N}} J(\mu, \nu) = J^* = \sup_{\nu \in \mathcal{N}} \inf_{\mu \in \mathcal{M}} J(\mu, \nu) =: \underline{J}. \tag{2.11b}$$

Here $\bar{J}$ and $\underline{J}$ are the *upper value* and the *lower value*, respectively, and generally we have the inequality $\bar{J} \geq \underline{J}$, as discussed in Section 2.1 in the context of static games. Only when they are equal, as in (2.11b), that the value J^* of the game is defined. ◊

In single player optimization (equivalently, optimal control) problems, the method of *dynamic programming* provides an effective means of obtaining the optimal solution whenever it exists, by solving a sequence of static optimization problems in retrograde time. The counterpart of this in two-player zero-sum dynamic games with CLPS information pattern is a recursive equation that involves (again in retrograde time) the saddle-point solutions of static games. Such an equation provides a sufficient condition for the existence of a saddle point for a dynamic game, and it will play an important role in the developments of Chapters 3 and 6. It is known as *Isaacs'* equation, being the discrete-time version of a similar equation obtained by Rufus Isaacs in the early 1950's in the continuous time [49] (see Section 2.3):

$$\begin{aligned}
V_k(x) &= \min_{u \in R^{m_1}} \max_{w \in R^{m_2}} [g_k(f_k(x, u, w), u, w, x) \\
&\qquad + V_{k+1}(f_k(x, u, w))] \\
&= \max_{w \in R^{m_2}} \min_{u \in R^{m_1}} [g_k(f_k(x, u, w), u, w, x) \\
&\qquad + V_{k+1}(f_k(x, u, w))]
\end{aligned}$$

$$\begin{aligned} &= g_k(f_k(x,\mu_k^*(x),\nu_k^*(x)),\mu_k^*(x),\nu_k^*(x),x) \\ &\quad + V_{k+1}(f_k(x,\mu_k^*(x),\nu_k^*(x))) \; ; \end{aligned} \quad (2.12)$$

$V_{K+1}(x) \equiv 0$.

We now have the following sufficiency result, whose proof can be found in ([17]; Chapter 6).

Theorem 2.4. *Let there exist a function $V_k(\cdot), k \geq 1$, and two policies $\mu^* \in \mathcal{M}_{\text{CLPS}}, \nu^* \in \mathcal{N}_{\text{CLPS}}$, generated by (2.12). Then, the pair (μ^*,ν^*) provides a saddle point for the discrete-time dynamic game formulated in this section, with the corresponding (saddle-point) value being given by $V_1(x_1)$.* ◇

Note that the policies $(\mu^* = \mu_{[1,K]}^*, \nu^* = \nu_{[1,K]}^*)$ are generated recursively in retrograde time, and for each $k \in [1, K]$ the computation of the pair (μ_k^*, ν_k^*) involves the saddle-point solution of a static game with (initial) state x_k. This makes these policies dependent only on the current value of the state, which are therefore called *feedback* policies. A second point to note in (2.12) is the requirement that every static game encountered in the recursion have a (pure-strategy) saddle point. If this requirement does not hold, then the first and second lines of (2.12) lead to different values, say $\bar{V}_k$ and $\underline{V}_k$, respectively, with $\bar{V}_1(x_1)$ corresponding to the *upper value* and $\underline{V}_1(x_1)$ to the *lower value* of the dynamic game, assuming that these values are bounded [15].

Even if the sufficiency result of Theorem 2.4 holds, this does not necessarily imply that all saddle-point policies of the dynamic game under the CLPS information pattern are generated by (2.12). In the characterization of all saddle-point policies of a given dynamic game, or in the generation of saddle points under different information patterns, the following notion of a *representation of a strategy* (see, [17]; Chapter 5) proves to be very useful, as it will also become evident from the analyses of the chapters to follow.

Definition 2.2. *For a given two-player dynamic game with policy spaces $\mathcal{M}$ and $\mathcal{N}$ (for Players 1 and 2, respectively), let $(\mu \in \mathcal{M}, \nu \in \mathcal{N})$ and $(\tilde{\mu} \in \mathcal{M}, \tilde{\nu} \in \mathcal{N})$ be two pairs of policies. These are representations of each other if*

(i) they both generate the same unique state trajectory, say $\tilde{x}_{[1,K+1]}$, and

(ii) they admit the same pair of open-loop values on this trajectory, i.e.,

$$\mu_k(\tilde{x}_{[1,k]}) = \tilde{\mu}_k(\tilde{x}_{[1,k]}),$$
$$\nu_k(\tilde{x}_{[1,k]}) = \tilde{\nu}_k(\tilde{x}_{[1,k]}); \quad \forall k \in [1, K].$$

◇

Clearly, under dynamic information patterns, a given pair of policies will have infinitely many representations; furthermore, if this pair happens to provide a saddle-point solution, then every representation will be a candidate saddle-point solution for the game, as implied by the following useful result.

Theorem 2.5. *For a given dynamic game, let $(\mu^* \in \mathcal{M}, \nu^* \in \mathcal{N})$ and $(\tilde{\mu} \in \mathcal{M}, \tilde{\nu} \in \mathcal{N})$ be two pairs of saddle-point policies. Then, necessarily, they are representations of each other.*

Proof. It follows from the ordered interchangeability property of multiple saddle points (cf. Property 2.1) that $(\tilde{\mu}, \nu^*)$ is also a saddle-point pair. This implies that, with $\nu = \nu^*$ fixed, both μ^* and $\tilde{\mu}$ (globally) minimize the game performance index (i.e. $J(\mu, \nu^*)$) over $\mathcal{M}$, and hence are representations of each other. A symmetrical argument shows that ν^* and $\tilde{\nu}$ are also representations of each other, thus completing the proof. ◇

Faced with the possibility of existence of multiple saddle-point equilibria in dynamic games when (at least one of) the players have access to dynamic (such as CLPS, or CLIS) information,[2] a natural question to ask is whether one can *refine* the notion of a saddle-point solution further, so as to ensure unicity of equilibria. Two such refinement schemes, which we will occasionally refer to in the following chapters, are *strong time consistency* and *(asymptotic) noise insensitivity* [7]. We call a (saddle-point) solution "strongly time consistent," if it provides a solution to any truncated version of the original game, regardless of the values of the new initial states.[3] More precisely,

Definition 2.3: Strong time consistency. *From the original game defined on the time interval* $[1, K]$, *construct a new game on a shorter time interval* $[\ell, K]$, *by setting* $\mu_{[0,\ell-1]} = \alpha_{[0,\ell-1]}$, $\nu_{[0,\ell-1]} = \beta_{[0,\ell-1]}$, *where* $\alpha_{[0,\ell-1]}$ *and* $\beta_{[0,\ell-1]}$ *are fixed but arbitrarily chosen. Let* $(\mu^* \in \mathcal{M}_{\text{CLPS}}, \nu^* \in \mathcal{N}_{\text{CLPS}})$ *be a saddle-point solution for the original game. Then, it is* "*strongly time consistent,*" *if the pair* $(\mu^*_{[\ell,K]}, \nu^*_{[\ell,K]})$ *is a saddle-point solution of the new game (on the interval* $[\ell, K]$), *regardless of the choices for* $\alpha_{[0,\ell-1]}, \beta_{[0,\ell-1]}$, *and for every* ℓ, $2 \leq \ell \leq K$. ◊

Clearly, every solution generated by the recursion (2.12) satisfies (by construction) this additional requirement, and is therefore strongly time consistent. Because of this special feature, such a closed-loop solution is also called a *feedback* saddle-point solution. A feedback solution is not necessarily unique, unless every static game encountered in (2.12) admits a unique saddle-point solution.

The second refinement scheme, "noise insensitivity," or its weaker version, "asymptotic noise insensitivity", refers to the robustness of the saddle-point solution to additive stochastic perturbations in the state dynamics,

---

[2] All these equilibria lead necessarily to the same saddle-point value (in view of Theorem 2.5), but not necessarily to the same existence conditions – an important point which will become clear in subsequent chapters.

[3] A weaker notion is that of *weak time consistency*, where the truncated game is assumed to have an initial state on the equilibrium trajectory of the original game [7]. Every saddle-point solution is necessarily weakly time consistent.

modeled by zero-mean white noise processes. Such a "stochastically perturbed" game will have its state dynamics (2.7a) replaced by

$$x_{k+1} = f_k(x_k, u_k, w_k) + \theta_k, \quad k \in [1, K],$$

where $\{\theta_k, k \in [1, K]\}$ is a sequence of n-dimensional independent zero-mean random vectors, with probability distribution $\mathcal{P}^\theta$. Furthermore, the performance index of the perturbed game will be given by the expected value of (2.8), with the expectation taken with respect to $\mathcal{P}^\theta$. We are now in a position to introduce the notion of (*asymptotic*) *noise insensitivity*.

Definition 2.4: Noise insensitivity. *A saddle-point solution of the unperturbed dynamic game under a given information pattern is noise insensitive if it also provides a saddle point for the stochastically perturbed game introduced above, for every (zero-mean, independent) probability distribution $\mathcal{P}^\theta$. It is asymptotically noise insensitive if it can be obtained as the limit of every saddle-point solution of the perturbed game, as $\mathcal{P}^\theta$ uniformly converges[4] to the one-point measure concentrated at $\theta = 0$.* ◇

Another appealing feature of the feedback saddle-point solution generated by *Isaacs*' equation (2.12) is that it is asymptotically noise insensitive. If the state dynamics are linear, and the objective function is quadratic in the three variables, as in (1.6a)-(1.7a), the feedback solution is in fact noise insensitive. This important property will be further elucidated on in the next chapter.

---

[4] Here, the convergence is defined with respect to the weak topology on the underlying sample space (a Polish space); see, for example, [48]. pp. 25-29.

2.3 Continuous-Time Dynamic Games

In the continuous-time formulation, (2.7a), (2.7b) and (2.8) are replaced, respectively, by

$$\frac{d}{dt}x(t) =: \dot{x} = f(t; x(t), u(t), w(t)), \quad t \geq 0 \tag{2.13a}$$

$$y(t) = h(t; x(t), w(t)), \quad t \geq 0 \tag{2.13b}$$

$$L(u, w) = q(x(t_f)) + \int_0^{t_f} g(t; x(t), u(t), w(t))dt, \tag{2.14}$$

where t_f is the terminal time, and the initial state x_0 is taken to be fixed (at this point) and known to both players.

Here we introduce five types of information structure, listed below:[5]

(i) *Closed-loop perfect state* (CLPS):

$$u(t) = \mu(t; x(s), s \leq t), \quad t \geq 0. \tag{2.15a}$$

Here we have to impose some additional (other than measurability) conditions on μ so that the state differential equation (2.13a) admits a unique solution. One such condition is to assume that $\mu(\cdot)$ is Lipschitz-continuous in x. We will not discuss these conditions in detail here, because it will turn out (see, Chapters 4 and 8) that they are not explicitly used in the derivation of the relevant saddle-point solution in the linear-quadratic game of real interest to us. Nevertheless, let us introduce the policy space $\mathcal{M}_{\text{CLPS}}$ to denote the general class of smooth closed-loop policies.

(ii) *Sampled-data perfect state* (SDPS): Here we have a partitioning of the time interval $[0, t_f]$, as

$$0 < t_1 < t_2 < \ldots < t_K < t_f$$

where t_k, $k \in \{1, \ldots, K\}$, is a sampling point. The controller has access to the values of the state only at the past (and present, if any)

---
[5] Again, we list them here only with respect to Player 1.

sampling points, i.e.,

$$u(t) = \mu(t; x(t_k), \ldots, x(t_1), x_0),$$
$$t_k \leq t < t_{k+1}, \quad k \in \{0, 1, \ldots, K\} \quad (2.15b)$$

where $\mu(\cdot)$ is piecewise continuous in t, and measurable in $x(t_k)$, $x(t_{k-1}), \ldots, x_0$. We denote the class of such policies by $\mathcal{M}_{\text{SDPS}}$.

(iii) *Delayed perfect state* (DPS): The controller has access to the state variable with a (constant) delay of $\theta > 0$ time units. Hence, admissible controllers are of the form:

$$\begin{aligned} u(t) &= \mu(t; x_{[0, t-\theta]}), \quad t \geq \theta \\ &= \mu(t; x_0), \quad 0 \leq t < \theta, \end{aligned} \quad (2.15c)$$

where $\mu(\cdot)$ is piecewise continuous in t, and Lipschitz-continuous in x. We denote the class of such control policies by $\mathcal{M}_{\theta D}$.

(iv) *Open-loop* (OL):

$$u(t) = \mu(t; x_0), \quad t \geq 0 \quad (2.15d)$$

where $\mu(\cdot)$ is piecewise continuous in t and measurable in x_0. The corresponding policy space is denoted by $\mathcal{M}_{\text{OL}}$.

(v) *Closed-loop imperfect state* (CLIS): The control is allowed to depend on the current as well as the entire past values of the measurement vector, i.e.,

$$u(t) = \mu(t; y(s), s \leq t), \quad t \geq 0, \quad (2.15e)$$

where the dependence on the initial state, whose value is known, has been suppressed. Here again we have to impose some regularity and growth conditions on μ so that the state differential equation admits a unique solution. We let $\mathcal{M}_{\text{CLIS}}$ denote the set of all such (Borel measurable) policies. Sampled-data and delayed versions of this CLIS information pattern will also be used in this book; but we postpone their introduction until Chapter 5.

The normal form for a continuous-time dynamic (differential) game can be defined as in (2.10), with Definition 2.1 being equally valid here for a saddle point. The counterpart of the recursion (2.12) is now the following continuous-time *Isaacs'* equation, which is a generalization of the Hamilton-Jacobi-Bellman partial differential equation that provides a sufficient condition in optimal control:

$$-\frac{\partial V(t;x)}{\partial t} = \min_{u \in R^{m_1}} \max_{w \in R^{m_2}} \left[\frac{\partial V(t;x)}{\partial x} f(t;x,u,w) + g(t;x,u,w)\right]$$

$$= \max_{w \in R^{m_2}} \min_{u \in R^{m_1}} \left[\frac{\partial V(t;x)}{\partial x} f(t;x,u,w) + g(t;x,u,w)\right]$$

$$= \frac{\partial V(t;x)}{\partial x} f(t;x,\mu^*(t;x),\nu^*(t;x))$$

$$+ g(t;x,\mu^*(t;x),\nu^*(t;x)) ;$$

$$V(t_f;x) \equiv q(x) .$$

(2.16)

The following is now the counterpart of Theorem 2.4, in the continuous time, whose proof can be found in ([17]; Chapter 8).

Theorem 2.6. *Let there exist a continuously differentiable function $V(t;x)$ on $[0,t_f] \times I\!R^n$, satisfying the partial differential equation (2.16), and two policies $\mu^* \in \mathcal{M}_{\text{CLPS}}$, $\nu^* \in \mathcal{N}_{\text{CLPS}}$, as constructed above. Then, the pair (μ^*,ν^*) provides a saddle point for the (continuous-time) differential game formulated in this section, with the corresponding (saddle-point) value being given by $V(0;x_0)$.* ◇

The notion of a *representation* of a strategy in continuous time can be introduced exactly as in Definition 2.2. Furthermore, the notions of *strong time consistency* and *(asymptotic) noise insensitivity* introduced in the discrete time in Section 2.2 find natural counterparts here. For the former, we simply replace the discrete time interval $[1,K]$ with the continuous interval $[0,t_f]$:

Definition 2.5: Strong time consistency. From the original game defined on the time interval $[0, t_f]$, construct a new game on a shorter time interval $[\ell, t_f]$, by setting $\mu_{[0,\ell)} = \alpha_{[0,\ell)}$, $\nu_{[0,\ell)} = \beta_{[0,\ell)}$, where $\alpha_{[0,\ell)}$ and $\beta_{[0,\ell)}$ are fixed but arbitrarily chosen. Let $(\mu^* \in \mathcal{M}_{\text{CLPS}}, \nu^* \in \mathcal{N}_{\text{CLPS}})$ be a saddle-point solution for the original game. Then, it is "strongly time consistent," if the pair $(\mu^*_{[\ell,t_f]}, \nu^*_{[\ell,t_f]})$ is a saddle-point solution of the new game (on the interval $[\ell, t_f]$), regardless of the admissible choices for $\alpha_{[0,\ell)}, \beta_{[0,\ell)}$, and for every $\ell \in (0, t_f)$. ◇

Every solution generated by *Isaacs'* equation (2.16) is strongly time consistent, and it is called (as in the discrete-time case) a *feedback* saddle-point solution.

Noise insensitivity can also be defined as in Definition 2.4, with the stochastic perturbation now modeled as a zero-mean independent-increment process. For the linear-quadratic differential game, it suffices to choose this as a standard Wiener process, with respect to which the feedback saddle-point solution is also noise insensitive. For details, the reader is referred to [[17], chapter 6].

Chapter 3

The Discrete-Time Minimax Design Problem With Perfect State Measurements

3.1 Introduction

In this chapter, we study the discrete-time minimax controller design problem, as formulated by (1.6)-(1.7), when the controller is allowed to have perfect access to the system state, either without or with one step delay. We first consider the case when the controlled output is a concatenation of the system state and the current value of the control, and the initial state is zero; that is, the system dynamics and the performance index are (without any loss of generality in this class):

$$x_{k+1} = A_k x_k + B_k u_k + D_k w_k, \quad x_1 = 0; \quad k \in [1, K] \qquad (3.1)$$

$$L(u, w) = |x_{K+1}|^2_{Q_f} + \sum_{k=1}^{K} \left\{ |x_k|^2_{Q_k} + |u_k|^2 \right\}, \qquad (3.2a)$$

$$Q_f \geq 0; \quad Q_k \geq 0, \quad k = 1, 2, \ldots$$

Note that in terms of the notation of (1.6b), we have

$$\widehat{H}_k = I, \quad \widehat{G}_{k-1} = 0, \quad \widehat{F}_{k-1} = 0, \quad H'_k H_k = Q_k, \quad H'_k G_k = 0, \quad G'_k G_k = I.$$

For this class of problems, we study in Section 2 the boundedness of the upper value, and the existence and characterization of the saddle-point solution of the associated soft-constrained linear-quadratic game with objective function (1.7b), which in this case is written as:

$$L_\gamma(u,w) = |x_{K+1}|^2_{Q_f} + \sum_{k=1}^{K} \left\{ |x_k|^2_{Q_k} + |u_k|^2 - \gamma^2 |w_k|^2 \right\}, \qquad (3.2b)$$

$$Q_f \geq 0; \quad Q_k \geq 0, \quad k = 1, 2, \ldots$$

We will alternatively use the notation $J_\gamma(\mu, w)$, for L_γ, as introduced in Section 1.3, when the control u is generated by a particular policy $\mu \in \mathcal{M}$, compatible with the given information pattern. We study this soft-constrained dynamic game under three different information patterns; namely, open-loop, closed-loop perfect state (CLPS), and closed-loop one-step delay (CLD), which were introduced in Section 2.2. We then use, in Section 3.3, the solutions of the soft-constrained games to construct optimal (minimax) controllers for the disturbance attenuation problem, under the CLPS and CLD information patterns. We also study the saddle-point property of this optimal controller in the context of the original hard-constrained disturbance attenuation problem, with the disturbance taken as a random sequence. Section 3.4 develops the counterparts of these results for the time-invariant infinite-horizon problem, and Section 3.5 extends them to the more general class where the controlled output is given by (1.6b), and the initial state x_1 is unknown. Section 3.6 discusses some extensions to nonlinear systems, and the chapter concludes with Section 3.7 which provides a summary of the main results.

3.2 The Soft-Constrained Linear-Quadratic Dynamic Game

We study the soft-constrained linear-quadratic dynamic game described by (3.1) and (3.2b), under three different information patterns, where we take the initial state x_1 to be known, but not necessarily zero, as in Section 2.2.

3.2.1 Open-loop information structure for both players.

The procedure to obtain the saddle-point solution of this game is first to fix the open-loop policy of Player 2, say $\bar{w}_{[1,K]}$, and minimize $L_\gamma(u, \bar{w}_{[1,K]})$ given by (3.2b) with respect to $u = u_{[1,K]}$. From linear-quadratic optimal control theory (see, e.g., [56]) it is well-known that the solution exists and is unique (since L_γ is strictly convex in u), and is characterized in terms of the standard discrete-time Riccati equation. Furthermore, the optimal u (to be denoted u^*) is linear in $w_{[1,K]}$ and x_1, i.e., for some linear function $\ell^1(\cdot)$, $u^*_{[1,K]} = \ell^1(\bar{w}_{[1,K]}, x_1)$. Now, conversely, if we fix the open-loop policy of Player 1, say $\bar{u}_{[1,K]}$, and maximize $J_\gamma(\bar{u}_{[1,K]}, w)$ with respect to $w = w_{[1,K]}$, the solution will again be linear, i.e., $w^*_{[1,K]} = \ell^2(\bar{u}_{[1,K]}, x_1)$, but existence and uniqueness will depend on whether L_γ is strictly concave in w or not. This requirement translates into the following condition:

Lemma 3.1. *The quadratic objective functional $L_\gamma(u, w)$ given by (3.2b), and under the state equation (3.1), is strictly concave in w for every open-loop policy u of Player 1, if, and only if,*

$$\gamma^2 I - D'_k S_{k+1} D_k > 0, \quad k \in [1, K] \tag{3.3a}$$

where the sequence S_{k+1}, $k \in [1, K]$, is generated by the Riccati equation

$$S_k = Q_k + A'_k S_{k+1} A_k + A'_k S_{k+1} D_k \left[\gamma^2 I - D'_k S_{k+1} D_k\right]^{-1} D'_k S_{k+1} A_k;$$

$$S_{K+1} = Q_f. \tag{3.3b}$$

◇

Under condition (3.3a), the quadratic open-loop game becomes strictly convex-concave, and it follows from Theorem 2.3[1] that it admits a unique

---
[1] Here, even though the sets U and W are not compact, it follows from the quadratic nature of the objective function that u and w can be restricted to closed and bounded (hence compact) subsets of finite-dimensional spaces, without affecting the saddle-point solution. Hence, Theorem 2.3 can be used.

(pure-strategy) saddle point – – which has to be the unique fixed point of the linear mappings:

$$u_{[1,K]} = \ell^1\left(w_{[1,K]}, x_1\right); w_{[1,K]} = \ell^2\left(u_{[1,K]}, x_1\right).$$

Certain manipulations, details of which can be found in ([17], p. 248), lead to the conclusion that this unique fixed point is characterized in terms of a matrix sequence generated by another (other than (3.3b)) discrete-time Riccati equation. The result is given in the following theorem:

Theorem 3.1. *For the discrete-time linear-quadratic zero-sum dynamic game with open-loop information structure, let (3.3a) be satisfied and M_k, $k \in [1, K]$, be a sequence of matrices generated by*

$$M_k = Q_k + A_k' M_{k+1} \Lambda_k^{-1} A_k; \quad M_{K+1} = Q_f \qquad (3.4a)$$

where

$$\Lambda_k := I + \left(B_k B_k' - \gamma^{-2} D_k D_k'\right) M_{k+1}. \qquad (3.4b)$$

Then,

(i) Λ_k, $k \in [1, K]$, are invertible.

(ii) The game admits a unique saddle-point solution, given by

$$u_k^* = \mu_k^*(x_1) = -B_k' M_{k+1} \Lambda_k^{-1} A_k x_k^* \qquad (3.5a)$$

$$w_k^* = \nu_k^*(x_1) = \gamma^{-2} D_k' M_{k+1} \Lambda_k^{-1} A_k x_k^*, \quad k \in [1, K] \qquad (3.5b)$$

where $\{x_{k+1}^*, k \in [1, K]\}$ is the corresponding state trajectory, generated by

$$x_{k+1}^* = \Lambda_k^{-1} A_k x_k^*, \quad x_1^* = x_1. \qquad (3.5c)$$

(iii) The saddle-point value of the game is

$$L_\gamma^* = L_\gamma(u^*, w^*) = x_1' M_1 x_1. \qquad (3.6)$$

(iv) If, however, the matrix in (3.3a) has at least one negative eigenvalue, then the upper value becomes unbounded. ◇

Before closing this subsection, we note that if the matrix sequence generated by (3.4a)-(3.4b) is invertible, then (3.4a) can be written in the more appealing (symmetric) form

$$M_k = Q_k + A'_k(M_{k+1}^{-1} + B_k B'_k - \gamma^{-2} D_k D'_k)^{-1} A_k; \quad M_{K+1} = Q_f. \quad (3.4')$$

A sufficient condition for invertibility (and hence positive definiteness) of M_k, $k \in [1, K]$, is that $(A'_k \; H'_k)'$, $k \in [1, K]$ be injective, that is rank $(A'_k \; H'_k) = n$, $k \in [1, K]$, and Q_f be positive definite.

3.2.2 Closed-loop perfect state information for both players.

When players have access to closed-loop state information (with memory), then, as discussed in Section 2.2, we cannot expect the saddle-point solution to be unique. There will in general exist a multiplicity of saddle-point equilibria – – each one leading to the same value (see Theorem 2.5), but not necessarily requiring the same existence conditions. To ensure unicity of equilibrium, one can bring in one of the refinement schemes introduced in Section 2.2, but more importantly, because of the intended application of this result, we have to make sure that the solution arrived at requires the least stringent condition on the positive parameter γ, among all saddle-point solutions to the problem. Both these considerations in fact lead to the same (*feedback* saddle-point[2]) solution, which is given in the following theorem.

Theorem 3.2. *For the two-person zero-sum dynamic game with closed-loop perfect state information pattern, and with a fixed $\gamma > 0$,*

---
[2]For this terminology, see Section 2.2.

(i) There exists a unique feedback saddle-point solution if, and only if,

$$\Xi_k := \gamma^2 I - D_k' M_{k+1} D_k > 0, \quad k \in [1, K] \tag{3.7}$$

where the sequence of nonnegative definite matrices M_{k+1}, $k \in [1, K]$, is generated by (3.4a).

(ii) Under condition (3.7), the matrices Λ_k, $k \in \mathcal{K}$, are invertible, and the unique feedback saddle-point policies are

$$u_k^* = \mu_k^*(x_k) = -B_k' M_{k+1} \Lambda_k^{-1} A_k x_k \tag{3.8}$$

$$w_k^* = \nu_k^*(x_k) = \gamma^{-2} D_k' M_{k+1} \Lambda_k^{-1} A_k x_k, \quad k \in [1, K], \tag{3.9}$$

with the corresponding unique state trajectory generated by the difference equation

$$x_{k+1}^* = \Phi_k^* x_k^* \equiv \Lambda_k^{-1} A_k x_k^*, \quad x_1^* = x_1, \tag{3.10}$$

and the saddle-point value is the same as (3.6), that is

$$J_\gamma^* = x_1' M_1 x_1, \tag{3.11}$$

(iii) If the matrix Ξ_k in (3.7) has a negative eigenvalue for some $k \in [1, K]$, then the game does not admit a saddle point under any information structure, and its upper value becomes unbounded.

Proof. Parts (i) and (ii) of the Theorem follow by showing that the given saddle-point policies uniquely solve the Isaacs' equation (2.12) for this linear-quadratic game, with the corresponding value function being

$$V_k(x) = x' M_k x, \quad k \in [1, K]. \tag{*}$$

For each k, existence of a unique saddle point to the corresponding static game in (2.12) is guaranteed by the positive definiteness of Ξ_k, and if there is some $\bar{k} \in [1, K]$ such that $\Xi_{\bar{k}}$ has a negative eigenvalue, then the corresponding static game in the sequence does not admit a saddle point, and being a quadratic game this implies that the upper value of the game (which then is different from the lower value) is unbounded. If Ξ_k has a zero (but no negative) eigenvalue, then whether the corresponding game admits a saddle point or not depends on the precise value of the initial state x_1, and in particular if $x_1 = 0$ one can allow Ξ_k to have a zero eigenvalue and still preserve the existence of a saddle point. Since the "zero-eigenvalue case" can be recovered as the limit of the "positive-eigenvalue case", we will henceforth not address the former. Some matrix manipulations, details of which can be found in ([17], pp. 247-257), show that condition (3.7) actually implies the nonsingularity of Λ_k.

We now prove part (iii), which is the *necessity* part of the result. Contrary to the statement of the Theorem, suppose that there exists a policy for Player 1, say $\{\hat{\mu}_k \in \mathcal{M}_k, k \in [1, K]\}$, under which the cost function J_γ is bounded for all $w_{[1,K]}$, even though Ξ_k has a negative eigenvalue for some $k \in [1, K]$; clearly, this policy cannot be the feedback policy (3.8). Let $\bar{k}$ be the largest integer in $[1, K]$ for which Ξ_k has a negative eigenvalue, and $\Xi_{\bar{k}+1} > 0$. Furthermore, let a policy for Player 2 be chosen as

$$\hat{\nu}_k(x_{[1,k]}) = \begin{cases} 0 & \text{for } k < \bar{k} \\ \hat{w}_k & \text{for } k = \bar{k} \\ \nu_k^*(x_k) & \text{for } k > \bar{k} \end{cases}$$

where $\hat{w}_{\bar{k}}$ is (at this point) an arbitrary element of $\mathbb{R}^{m_2}$, and ν_k^* is as defined by (3.9). Denote the state trajectory corresponding to the pair $(\hat{\mu}_{[1,K]}, \hat{\nu}_{[1,K]})$ by $\hat{x}_{[1,K]}$, and the corresponding *open-loop values* of the two policies by the two players by $\hat{u}_{[1,K]}$ and $\hat{w}_{[1,K]}$. Finally, we introduce the notation $J^{(k)}(u_{[k,K]}, w_{[k,K]}; \hat{x}_k)$ to denote the kernel of a linear-quadratic game, formulated in exactly the same way as the original game, with the only difference being that it is defined on the subinterval $[k, K]$ and has the initial state $\hat{x}_k$. Then, we have the following sequence of equalities and inequalities:

$$\sup_{w_{[1,K]}} J_\gamma(\hat{\mu}_{[1,K]}, w_{[1,K]}) \geq J(\hat{\mu}_{[1,K]}, \hat{\nu}_{[1,K]})$$

$$= J^{(\bar{k}+1)}(\hat{\mu}_{[\bar{k}+1,K]}, \nu^*_{[\bar{k}+1,K]}; \hat{x}_{\bar{k}+1}) + -\gamma^2|\hat{w}_{\bar{k}}|^2 + \sum_{k=1}^{\bar{k}}\{|\hat{x}_k|^2_{Q_k} + |\hat{u}_k|^2\}$$

$$\geq J^{(\bar{k}+1)}(\mu^*_{[\bar{k}+1,K]}, \nu^*_{[\bar{k}+1,K]}; \hat{x}_{\bar{k}+1}) + |\hat{u}_{\bar{k}}|^2 - \gamma^2|\hat{w}_{\bar{k}}|^2 + c_{\bar{k}}$$

$$= |\hat{x}_{[\bar{k}+1]}|^2_{M_{\bar{k}+1}} + |\hat{u}_{\bar{k}}|^2 - \gamma^2|\hat{w}_{\bar{k}}|^2 + c_{\bar{k}}$$

where $c_{\bar{k}}$ is some nonnegative constant (determined only by the initial state x_1 and Player 1's control $\hat{u}_{[1,\bar{k}-1]}$). In the above, the inequality in the first line follows because $\hat{\nu}$ is not necessarily a maximizing policy for Player 2; the equality of the next line follows because ν^* and $\hat{\nu}$ agree (by construction) on the subinterval $[\bar{k}+1, K]$; the inequality of the third line follows from the fact that the pair $(\mu^*_{[\bar{k}+1,K]}, \nu^*_{[\bar{k}+1,K]})$ provides a saddle-point solution to the truncated game with kernel $J^{(\bar{k}+1)}$ (this is true because the feedback saddle-point solution of the original game is *strongly time consistent*); and finally the equality of the last line follows from the expression (*) given above for the value function of the game. Since, by hypothesis, $\Xi_{\bar{k}+1}$ has a negative eigenvalue, the function

$$\hat{x}'_{\bar{k}+1} M_{\bar{k}+1} \hat{x}_{\bar{k}+1} - \gamma^2 \hat{w}'_{\bar{k}} \hat{w}_{\bar{k}}$$

is not bounded above (in $\hat{w}_{\bar{k}}$), and since the choice of $\hat{w}_{\bar{k}}$ was arbitrary this shows that the kernel $J(\hat{\mu}_{[1,K]}, w_{[1,K]})$ can be made arbitrarily large — a contradiction to the initial hypothesis that the upper value of the game was bounded. ◇

Several remarks are now in order, to place the various statements of the theorem above into proper perspective.

Remark 3.1. An important implication of the last statement of the theorem, just proved, is the property that under the closed-loop information pattern the feedback saddle-point solution requires the least stringent existence conditions; in other words, if the matrix in (3.7) has a negative eigenvalue for some $k \in [1, K]$, then a saddle point will not exist (and the upper value will be unbounded), even if Player 1's policy is allowed to depend on past values of the state. It then follows from the *ordered interchangeability* property (Property 2.1), and particularly Theorem 2.5, that saddle-point solutions under other information structures (provided that these saddle points exist) can be constructed from the feedback saddle-point solution (3.8)-(3.9) on the state trajectory (3.10). In other words, every other saddle-point solution of the original game is a *representation* (cf. Definition 2.2) of the feedback solution on the saddle-point trajectory. In mathematical terms, any closed-loop policy $u_k = \mu_k(x_{[1,k]})$, $k \in [1, K]$, satisfying the following two side conditions, constitutes a saddle-point solution along with (3.9):

(i) $\quad \mu_k(x^*_{[1,k]}) = -B'_k M_{k+1} \Lambda_k^{-1} A_k x^*_k, \quad k \in [1, K].$ \hfill (3.12)

(ii) $\quad J_\gamma(\mu_{[1,K]}, w_{[1,K]})$ is strictly concave in $w_{[1,K]}$.

Side condition (ii) above leads, in general, (depending on $\mu_k, k \in [1, K]$) to a more (and never less) stringent condition than (3.7), an example in point being the open-loop solution $u^*_k = \mu^*_k(x^*_k)$, $k \in [1, K]$, given by (3.5a), for which the "strict concavity" condition is (3.3a), which can directly be shown to be more restrictive than condition (3.7) (see, [17], p.258). ◇

Remark 3.2. Another point worth noting is that if the information available to Player 2 is not closed-loop (or feedback), but rather open-loop, the statement of Theorem 3.2 remains equally valid, with only x_k in (3.9) replaced by x^*_k which is generated by (3.10). In fact, quite parallel to the discussion above, under a general closed-loop pattern for the maximizer, any closed-loop policy $w_k = \nu(x_{[1,k]})$, $k \in [1, K]$, for the maximizer, that satisfies the following two side conditions, will be in saddle-point equilibrium with (3.8):

(i) $\quad \nu_k(x^*_{[1,k]}) = \gamma^{-2} D'_k M_{k+1} \Lambda_k^{-1} A_k x^*_k, \quad k \in \mathcal{K}$ \hfill (3.13) i.e., ν should be a *representation* of ν^* on the equilibrium trajectory (3.10).

(ii) $J_\gamma(u_{[1,K]}, \nu_{[1,K]})$ is strictly convex in $u_{[1,K]}$. ◇

Another variation of the basic game problem covered by Theorem 3.2 is the one where Player 1 has the expanded information structure where he is allowed to observe (in addition to the state) the current value of the control of Player 2. Then the condition of Theorem 3.2 can be further relaxed, since (in essence) this extra information gives an advantage to the minimizer. The problem then is a special type of a *Feedback Stackelberg Game* ([17], p. 314), where the derivation of the saddle-point solution can be done recursively, with the difference from the one of Theorem 3.2 being that now in the *Isaacs'* equation (2.12) the *maxmin value* of each static game is considered (instead of the *minimax* or *saddle-point value*). The resulting solution is both *strongly time consistent* and *noise insensitive*, as given in the following theorem:

Theorem 3.3. *Consider the linear-quadratic dynamic game of Theorem 3.2, but now with Player 1's control allowed to depend on the current value of the action of Player 2 (in addition to its dependence on the state). Then,*

(i) *The dynamic game admits a unique strongly time consistent and noise insensitive saddle-point solution if, and only if, the following condition is satisfied:*

$$\gamma^2 I - D_k'(I + M_{k+1} B_k B_k')^{-1} M_{k+1} D_k > 0, \quad k \in [1, K], \quad (3.14)$$

which is equivalent to the condition that the matrix Λ_k defined by (3.4b) has only positive eigenvalues, for all $k \in [1, K]$.

(ii) *Under condition (3.14), the unique strongly time-consistent saddle-*

point policy of Player 1 is given by

$$\tilde{u}_k = \tilde{\mu}_k(x_k, w_k) = -(I + B'_k M_{k+1} B_k)^{-1} B'_k M_{k+1}(D_k w_k + A_k x_k),$$

$$k \in [1, K],$$

(3.15)

whereas that of Player 2 is still given by (3.9).

(iii) If the matrix in (3.14) has a negative eigenvalue for at least one $k \in [1, K]$, then the upper value of the game under this asymmetric information (which is also the lower value) is unbounded.

(iv) Whenever condition (3.14) holds, the saddle-point value of the game under this asymmetric information is still given by (3.11).

Proof. The proof again follows from the recursion (2.12), where now we solve (for each static game) Player 1's control u_k as a function of w_k (and x_k). This readily leads to the policy (3.15), and substitution of this into the same expression yields (3.14) as the strict concavity condition. We again have $V_k(x) = x' M_k x$, which can be shown recursively. ◇

It is worth noting here that since condition (3.7) implies the nonsingularity of Λ_k, and not *vice versa*, the increase in information to Player 1 has led to a less stringent existence condition, as to be expected.

3.2.3 Closed-loop 1-step delay information for both players.

Another possible realistic information structure for Player 1 is the one where the control is allowed to depend on the state with a delay of one time unit. Endowing Player 2 also with a similar information structure, the permissible policies for the two players will now be

$$\mu_k(x_{[1,k-1]}), \quad k \in [1, K] \; ; \quad \nu_k(x_{[1,k-1]}), \quad k \in [1, K].$$

In this case, for the original deterministic game the saddle-point solution is again not unique, because of the redundancy in the dynamic information.

However, requiring the solution to be (additive) noise insensitive (cf. Definition 2.4) leads to a unique solution by using Theorem 3.2, Property 2.1 and Theorem 2.5, and a dynamic programming type argument. The solution is essentially a particular *representation* of the feedback solution (3.8)-(3.9) on the state trajectory (3.10). Toward deriving this representation, we first note that if the control policies were restricted to depend **only** on the most recently available (delayed) value of the state, the unique saddle-point solution would be given by (using the notation of Theorem 3.2):[3]

$$u_k = \hat{\mu}_k(x_{k-1}) = -B_k' M_{k+1} \Lambda_k^{-1} A_k \Lambda_{k-1}^{-1} A_{k-1} x_{k-1}, \quad k > 1$$

$$= -B_1' M_2 \Lambda_1^{-1} A_1 x_1, \quad k = 1$$

(3.16a)

$$w_k = \hat{\nu}_k(x_{k-1}) = \gamma^{-2} D_k' M_{k+1} \Lambda_k^{-1} A_k \Lambda_{k-1}^{-1} A_{k-1} x_{k-1}, \quad k > 1$$

$$= \gamma^{-2} D_1' M_2 \Lambda_1^{-1} A_1 x_1, \quad k = 1$$

(3.16b)

which is a valid solution provided that the cost function $J_\gamma(\hat{\mu}_{[1,K]}, w_{[1,K]})$ is strictly concave in $w_{[1,K]}$. It is not difficult to see that this is not a noise-insensitive solution. Now, to construct one (in fact the unique one) that is noise insensitive and requires the least stringent existence condition, we rewrite (3.8) as

$$u_k = -B_k' M_{k+1} \Lambda_k^{-1} A_k \underbrace{(A_{k-1} x_{k-1} + B_{k-1} u_{k-1} + D_{k-1} w_{k-1})}_{\xi_k}$$

where ξ_k will have to be expressed in terms of not only x_{k-1}, but also $x_{[1,k-2]}$, through u_{k-1} and w_{k-1}. Likewise, (3.9) is rewritten as

$$w_k = \gamma^{-2} D_k^T M_{k+1} \Lambda_k^{-1} A_k \xi_k$$

in view of which, for all $k \in [1, K]$,

$$\xi_{k+1} = A_k x_k - (B_k B_k' - \gamma^{-2} D_k D_k') M_{k+1} \Lambda_k^{-1} A_k \xi_k, \quad \xi_1 = x_1$$

$\Leftrightarrow$

$$\xi_{k+1} = A_k x_k + (\Lambda_k^{-1} - I) A_k \xi_k, \quad \xi_1 = x_1. \quad (3.17)$$

---
[3] This result follows readily from Remark 3.1.

This is an n-dimensional compensator, the output of which at stage k replaces x_k in (3.8)-(3.9). The condition for the resulting set of policies to be in saddle-point equilibrium is strict concavity of $J_\gamma(\mu_{[1,K]}, w_{[1,K]})$ in $w_{[1,K]}$ when $\mu_{[1,K]}$ is given by

$$\mu_k^*(\xi_k) = -B_k' M_{k+1} \Lambda_k^{-1} A_k \xi_k, \quad k \in [1, K] \tag{3.18}$$

Let $\zeta_k := (x_k', \xi_k')'$, which is generated by

$$\zeta_{k+1} = \tilde{A}_k \zeta_k + \tilde{D}_k w_k, \quad \zeta_1 = (x_1', x_1')' \tag{3.19a}$$

where

$$\tilde{A}_k := \begin{pmatrix} I & -B_k B_k' M_{k+1} \Lambda_k^{-1} \\ I & \Lambda_k^{-1} - I \end{pmatrix} A_k; \quad \tilde{D}_k := \begin{pmatrix} D_k \\ 0 \end{pmatrix}. \tag{3.19b}$$

In terms of ζ_k, $k \in [1, K]$, $J_\gamma(\mu^*, w)$ can be written as

$$J_\gamma(\mu^*, w) = \sum_{k=1}^{K} \left\{ |\zeta_{k+1}|^2_{\tilde{Q}_{k+1}} - \gamma^2 |w_k|^2 \right\} \tag{3.19c}$$

where

$$\tilde{Q}_{K+1} = \text{diag}(Q_f, 0);$$

$$\tilde{Q}_k = \text{diag}\left(Q_k, \, A_k'(\Lambda_k^{-1})' M_{k+1} B_k B_k' M_{k+1} \Lambda_k^{-1} A_k\right), \quad k \in [1, K]. \tag{3.19d}$$

The condition for strict concavity of (3.19c), in $w_{[1,K]}$, is

$$\gamma^2 I - \tilde{D}_k' \tilde{S}_{k+1} \tilde{D}_k > 0, \quad k \in [1, K] \tag{3.20a}$$

where $\tilde{S}_{k+1}$, $k \in [1, K]$, is generated by

$$\tilde{S}_k = \tilde{Q}_k + \tilde{A}_k' \tilde{S}_{k+1} \tilde{A}_k + \tilde{A}_k' \tilde{S}_{k+1} \tilde{D}_k \left[\gamma^2 I - \tilde{D}_k' \tilde{S}_{k+1} \tilde{D}_k\right]^{-1} \tilde{D}_k' \tilde{S}_{k+1} \tilde{A}_k;$$
$$\tilde{S}_{K+1} = \tilde{Q}_{K+1}.$$
$$\tag{3.20b}$$

We are now in a position to state the following theorem.

Discrete-Time Problem with Perfect State Measurements

Theorem 3.4. *For the linear-quadratic zero-sum dynamic game with closed-loop 1-step delayed state information for both players the following hold:*

(i) *Under condition (3.20a), there exists a unique noise-insensitive saddle-point solution, given by (3.8)-(3.9) with x_k replaced by ξ_k which is generated by (3.17). The saddle-point value is again given by (3.11).*

(ii) *If the matrix in (3.20a) has a negative eigenvalue for at least one $k \in [1, K]$, then the game does not admit a saddle point, and its upper value is unbounded under the given information.*

Proof. The saddle-point property follows Theorem 2.5, in lieu of the observation that the given pair is a particular representation of the feedback saddle-point solution (3.8)-(3.9) on the saddle-point trajectory, and the upper value is bounded under (3.20a), as discussed in the construction above. The noise insensitivity and uniqueness in this class follow from an alternative recursive derivation of the solution using a dynamic programming type argument similar to (2.12). The proof of part (ii), on the other hand, is a proof by contradiction, paralleling that of Theorem 3.2(iii). ◊

Note that, as compared with the closed-loop information case, here we need in addition (to (3.4a)) the solution of a ($2n$-dimensional) Riccati equation (3.20b), which is though needed only to check the condition of existence. It follows from Theorem 3.2(iii) that condition (3.20a) implies condition (3.7), so that under the former the matrix Λ_k is invertible for all $k \in [1, K]$.

Another point to note is that if Player 2 has access to full state information (and Player 1 still having 1-step delayed information) condition (3.20a) is still operative, and the saddle-point solution given in the Theorem remains a saddle point (though no longer noise insensitive). In this case also the given saddle-point solution requires the least stringent condition among the infinitely many saddle-point solutions to the problem.

3.2.4 An illustrative example

To illustrate the results of Theorems 3.2 - 3.4, we now consider a scalar version of the dynamic game problem, where the state dynamics are

$$x_{k+1} = x_k + u_k + w_k, \quad k = 1, 2, \ldots \tag{3.21a}$$

and the objective function is

$$L_\gamma = \sum_{k=1}^{K} [x_{k+1}^2 + u_k^2 - r w_k^2] \tag{3.21b}$$

where $r = \gamma^2 > 0$ is taken as a variable.

(i) With $K = 3$, condition (3.7) reads

$$r > 1, \quad r > 1 + \frac{1}{\sqrt{2}} \approx 1.7071, \quad r > 1.9275^4$$

which implies that the problem admits a unique saddle point in feedback (FB) policies if, and only if, $r > 1.9275$. It is possible to show that as $K \to \infty$, the various constraints on r become nested, converging to $r > 2$.

Again for $K = 3$, the minimax (saddle-point) controllers are

$$u_3 = \mu_3^*(x_3) = -\frac{r}{2r-1} x_3; \quad u_2 = \mu_2^*(x_2) = -\frac{r(3r-1)}{5r^2 - 5r + 1} x_2$$

$$u_1 = \mu_1^*(x_1) = -\frac{r(8r^2 - 6r + 1)}{13r^3 - 19r^2 + 8r - 1} x_1 \tag{3.22}$$

and the FP saddle-point policy for the maximizer is computed using the formula

$$w_k = \nu_k^*(x_k) = -\frac{1}{r} \mu_k^*(x_k). \tag{3.23}$$

The corresponding state trajectory is

$$x_2^* = \frac{5r^3 - 5r^2 + r}{13r^3 - 19r^2 + 8r - 1} x_1$$

$$x_3^* = \frac{r(2r-1)}{5r^2 - 5r + 1} x_2^*; \quad x_4^* = \frac{r}{2r-1} x_3^* \tag{3.24}$$

---
[4] This bound is the largest root of the polynomial $5s^3 - 13s^2 + 7s - 1$.

For comparison purposes, let us now determine the condition (on r) for the game to have a bounded lower value. In view of Theorem 3.3, and particularly condition (3.14), we have

$$r > \frac{1}{2}, \quad r > \frac{5+\sqrt{5}}{10} \approx 0.7236, \quad r > 0.8347^5$$

and hence the lower (maxmin) value of the game with $K = 3$ is unbounded if $r < 0.8347$, and it is bounded if $r > 0.8347$. this condition is clearly less stringent than that of boundedness of the upper (minimax) value.

(ii) Let $K = 2$. Then the set of all saddle-point minimax controllers in the linear class is given by

$$u_1 = \mu_1(x_1) = -\frac{r(3r-1)}{5r^2 - 5r + 1} x_1 \quad (3.25a)$$

$$u_2 = \mu_2(x_2) = -\frac{r}{2r-1} x_2 + P\left(x_2 - \frac{r(2r-1)}{5r^2 - 5r + 1} x_1\right) \quad (3.25b)$$

where P is a scalar satisfying the concavity condition (see Remark 3.1):

$$P^2 < \frac{(r-1)(2r^2 - 4r + 1)}{(2r-1)^2}. \quad (3.25c)$$

As r approaches the boundary of the region associated with the FB solution (i.e., $r \to 1 + \frac{1}{\sqrt{2}}$), the right-hand side of (3.26) goes to zero, implying that the FB solution of Theorem 3.2 indeed provides the least stringent condition on the parameter r.

(iii) To illustrate the result of Theorem 3.4, let us take $K = 4$, and compute condition (3.20a) to obtain:

$$r > 4.87659$$

while condition (3.7) in this case dictates

$$r > 1.98218 \ .$$

Note the degradation in the concavity condition due to loss in the information available to the players (particularly, Player 1). For comparison purposes, let us also compute the bound on r under the simpler (no-memory)

---
[5] This bound is the largest root of the polynomial $13s^3 - 19s^2 + 8s - 1$.

controller (3.17a):

$$r > 6.31768$$

which is (as expected) the most stringent of the three.

3.3 Solution to the Disturbance Attenuation Problem

3.3.1 General closed-loop information

We now return to the original discrete-time disturbance attenuation problem formulated in Sections 1.2 and 1.3, with performance index (3.2a) and CLPS information pattern, and apply the results of the previous section by following the general line of argument introduced in Section 1.2. Note that the objective here is to find a controller $\mu^* \in \mathcal{M}_{\text{CLPS}}$ such that

$$\gamma^* = \inf_{\mu \in \mathcal{M}} \ll T_\mu \gg = \ll T_{\mu^*} \gg \tag{3.26a}$$

where γ^* is the *optimal* (*minimax*) attenuation level, and

$$\ll T_\mu \gg := \sup_{w \in \mathcal{H}_w} \|T_\mu(w)\|/\|w\| \equiv \sup_{w \in \mathcal{H}_w} \{J(\mu, w)\}^{1/2}/\|w\| \tag{3.26b}$$

with J given by (3.2a). Toward obtaining the minimax controller μ^*, and the corresponding attenuation level γ^*, let us first fix γ such that condition (3.7) of Theorem 3.2 is satisfied. For this fixed value of γ let the feedback controller given by (3.8) be denoted by μ^γ, and the disturbance policy in (3.9) by ν^γ. Then, in view of (3.11),

$$J_\gamma(\mu^\gamma, w_{[1,K]}) \leq J_\gamma(\mu^\gamma, \nu^\gamma) = x_1' M_1 x_1, \tag{3.27a}$$

with the inequality holding for all $x_1 \in \mathbb{R}^n$, and all l^2 sequences $w_{[1,K]} := (w_1, ..., w_K)$ (i.e., $w \in \mathcal{H}_w$). Using the relationship (1.5) between J and J_γ, the inequality (3.27a) can equivalently be written as

$$J(\mu^\gamma, w_{[1,K]}) \leq x_1' M_1 x_1 + \gamma^2 \sum_{k=1}^{K} |w_k|^2, \tag{3.27b}$$

and taking $x_1 = 0$, we arrive at

$$J(\mu^\gamma, w_{[1,K]}) \leq \gamma^2 \sum_{k=1}^{K} |w_k|^2, \quad \forall w \in \mathcal{H}_w. \tag{3.28a}$$

Now, using the relationship between J and $\ll T_\mu \gg$ as given in (3.26b), we finally have:

$$\ll T_{\mu^\gamma} \gg \leq \gamma. \tag{3.28b}$$

Hence, given any γ satisfying condition (3.7) of Theorem 3.2, the control law $\mu^\gamma = \mu^\gamma_{[1,K]}$ given by (3.8), which depends on this value of γ, delivers the desired bound (3.28b). Introduce the set

$$\Gamma := \{\gamma > 0 : \text{condition (3.7) holds}\}$$

which is a nonempty open set, because the condition holds for γ sufficiently large, and if a particular γ is a permissible choice then $\gamma - \epsilon$ also is, for $\epsilon > 0$ sufficiently small. Let $\hat{\gamma}$ be the infimum of Γ, that is

$$\hat{\gamma} := \inf\{\gamma : \gamma \in \Gamma\}.$$

Clearly, by Theorem 3.2(iii), there does not exist any closed-loop controller which will deliver a bound in (3.28b) that is smaller than $\hat{\gamma}^2$, and hence

$$\gamma^* = \hat{\gamma},$$

where the former was defined by (1.2a) (equivalently (3.26a)). Now let the control gain in (3.8) be denoted by $G_k(\gamma)$, to explicitly show the dependence of the minimax control law on the parameter γ. We first have the following useful Lemma.

Lemma 3.2. *Let $\{\gamma^{(n)}, n = 1, 2, ...\}$ be any decreasing sequence of positive numbers, with limit point $\hat{\gamma}$. Then, for each $k \in \mathcal{K}$, the sequence of matrix functions $\{G_k(\gamma^{(n)}), n = 1, 2, ...\}$ is right-continuous at $\hat{\gamma}$.*

Proof. All we need to show is that the control gain $G_k(\gamma)$ is bounded as $\gamma \downarrow \hat{\gamma}$. For this, we need only consider the case where $\Lambda_k(\hat{\gamma})$ is singular for at least one $k \in \mathcal{K}$, say $\bar{k}$, which will occur only if

$$\hat{\gamma}^2 I - D_k' M_{k+1} D_k \geq 0, \qquad k = \bar{k}, \tag{3.29}$$

with at least one zero eigenvalue. Now, it follows from (3.4b) and (3.7) that

$$\text{Ker}(\Lambda_{\bar{k}}) = \text{Ker}(B_{\bar{k}}' M_{\bar{k}+1}) \cap \text{Ker}(I - \hat{\gamma}^{-2} D_{\bar{k}} D_{\bar{k}}' M_{\bar{k}+1}),$$

which actually holds not only at $\gamma = \hat{\gamma}$, but also for $\gamma > \hat{\gamma}$. This then implies that the matrix

$$B_{\bar{k}}' M_{\bar{k}+1} \Lambda_{\bar{k}}^{-1}$$

(with the inverse on $\Lambda_{\bar{k}}$ interpreted in a limiting sense) is bounded for all $\gamma > \hat{\gamma}$, and also in the limit as $\gamma \downarrow \hat{\gamma}$. This proves the right-continuity of $\{G_{\bar{k}}(\gamma)\}$ at $\gamma = \hat{\gamma}$. ◇

Now, we are in a position to present the following result.

Theorem 3.5. *Let $\hat{\gamma} \geq 0$ be the scalar defined above (as the infimum of Γ), and $\mu_{\hat{\gamma}}^* := \mu_{[1,K]}^*$ be the feedback control law (3.8) with $\gamma = \hat{\gamma}$. Then, for the discrete-time finite-horizon disturbance attenuation problem, with performance index (3.2a), and CLPS information pattern,*

$$\inf_{\mu \in \mathcal{M}_{\text{CLPS}}} \sup_{w \in \mathcal{H}_w} \frac{\|T_\mu(w)\|}{\|w\|} \equiv \inf_{\mu \in \mathcal{M}_{\text{CLPS}}} \ll T_\mu \gg = \ll T_{\mu_{\hat{\gamma}}^*} \gg = \hat{\gamma}^2; \tag{3.30}$$

that is, the feedback controller $\mu_{\hat{\gamma}}^*$ solves the disturbance attenuation problem, and the optimal attenuation level equals $\hat{\gamma}$.

Proof. The proof follows basically from Theorem 3.2, and the discussion leading to the Theorem. We provide here a formal proof for the *necessity* part of the result, which is that the value in (3.30) [which is the upper value of the disturbance attenuation game] cannot be smaller than $\hat{\gamma}$. Suppose that, to the contrary, the value is $\tilde{\gamma} < \hat{\gamma}$. Then, there exists a controller $\tilde{\mu}_{[1,K]} \in \mathcal{M}_{\text{CLPS}}$ such that for some $\epsilon > 0$ (sufficiently small, so that $\tilde{\gamma}^2 + \epsilon < \hat{\gamma}^2$)

$$\ll T_{\tilde{\mu}} \gg = \tilde{\gamma}^2 + \epsilon =: \tilde{\tilde{\gamma}}^2 < \hat{\gamma}^2,$$

which is equivalent to the inequality

$$J(\tilde{\mu}_{[1,K]}, w_{[1,K]}) - \tilde{\tilde{\gamma}}^2 \sum_{k=1}^{K} |w_k|^2 \le 0, \quad \text{for all } w_{[1,K]} \in \mathcal{H}_w$$

This, however, is impossible by Theorem 3.2(iii) (see also Remark 3.1), because for $\gamma < \hat{\gamma}$ the upper value of the associated *soft-constrained* game of Section 3.2.1 is unbounded. The fact that the bound is achieved under $\mu^*_{\hat{\gamma}}$ follows from the continuous dependence of the eigenvalues of a (symmetric) matrix on its entries.

Note that since the control policy space $\mathcal{M}_{\text{CLPS}}$ also includes nonlinear controllers, use of a nonlinear controller cannot improve upon the attenuation bound $\hat{\gamma}$. ◇

Remark 3.3. Two aspects of the solution presented above are worth emphasizing. The first is that in the discrete-time finite-horizon[6] disturbance attenuation problem with perfect state measurements the existence of an optimal controller is always guaranteed, while as we shall see in Chapter 4 this is not so for the continuous-time version of the problem. The second point is the cautionary remark that the theorem does not attribute any (pure-strategy) saddle point to the disturbance attenuation problem; in other words, there is no claim that "inf" and "sup" operations in (1.3) can be interchanged. Even though the related (soft-constrained) dynamic game problem of Section 3.2 admitted a pure-strategy saddle point, the disturbance attenuation problem in fact does not admit one. What one can show, however, is that the minimax controller $\mu^*_{\hat{\gamma}}$ is in equilibrium with

---

[6] We will shortly see that this is true also for the infinite-horizon case.

a "mixed" l^2 policy for the disturbance, which features a discrete distribution. A precise characterization of this worst case distribution is given later in Section 3.3.3, (see Proposition 3.1). To motivate this result, we will first obtain (in the sequel) the worst case distribution for the two and three stage examples treated earlier in Section 3.2.4. ◇

3.3.2 Illustrative example (continued)

Let us return to the example of Section 3.2.4, treated now as a disturbance attenuation problem, with the objective function (3.21b) replaced by

$$J = \sum_{k=1}^{K} [x_{k+1}^2 + u_k^2] \qquad (3.31)$$

and with the initial state taken to be $x_1 = 0$.

(i) With $K = 3$, the attenuation constant $\hat{\gamma}$ used in Theorem 3.5 is

$$\hat{\gamma} = \sqrt{1.9275} = 1.3883, \qquad (3.32)$$

leading to the unique feedback minimax control law (from (3.23))

$$\mu_3^*(x_3) = -0.675131 x_3, \quad \mu_2^*(x_2) = -0.927505 x_2, \quad \mu_1^* = 0. \qquad (3.33)$$

The corresponding mixed (worst-case) disturbance can obtained from (3.23) as follows: We first compute the limit of the open-loop saddle-point disturbance (obtained by taking $x_1 \neq 0$) as $\gamma \downarrow \hat{\gamma}$:

$$w_3 = 0.100616 x_1, \quad w_2 = 0.249647 x_1, \quad w_1 = 0.518807 x_1 \qquad (3.34)$$

This, of course, in itself is meaningless since for the problem at hand $x_1 = 0$. However, the ratios of w_3 and w_2 to w_1 are meaningful quantities, independent of x_1, and if we further normalize the total energy (i.e., $\|w\|^2$) to unity (without any loss of generality), we arrive at the following distribution of unit energy across stages:

$$w_3^+ = 0.172149, \quad w_2^+ = 0.427133, \quad w_1^+ = 0.887651. \qquad (3.35)$$

Then, the worst-case disturbance (under unit energy constraint) that is in saddle-point equilibrium with the minimax controller (3.33) is the mixed policy

$$w_k^* = \xi w_k^+, \quad k = 1, 2, 3, \tag{3.36}$$

where ξ is a discrete random variable taking the values $+1$ and -1 with equal probability $1/2$, and w_k^+ is given by (3.35). This is unique up to a normalization constant, which is determined by the energy bound b.

(ii) With $K = 2$, the value of γ^* is

$$\gamma^* = \sqrt{1 + (1/\sqrt{2})} \approx 1.30656, \tag{3.37}$$

and the corresponding (unique feedback) minimax controller is (from (3.27), with $P = 0$)

$$\mu_2^*(x_2) = -(1/\sqrt{2})x_2, \quad \mu_1^* = 0. \tag{3.38}$$

The worst-case disturbance, computed according to the procedure outlined above, is

$$w_1^* = \frac{\xi}{2}\sqrt{2 + \sqrt{2}}, \quad w_2^* = \frac{\xi}{2}\sqrt{2 - \sqrt{2}}, \tag{3.39}$$

where ξ is as defined above. For a different (more direct) method to obtain the solution above, see [5].

(iii) If we let $K \to \infty$ in the auxiliary game problem with perfect state information, we arrive at the stationary minimax controller

$$u_k = \mu^*(x_k) = (M(\gamma)/[1 + \frac{\gamma^2 - 1}{\gamma^2} M(\gamma)])x_k, \quad k = 1, 2, \ldots$$

provided that $\gamma > \sqrt{2}$. For the disturbance attenuation problem with infinite horizon, this leads to the attenuation constant $\gamma^* = \sqrt{2}$ and to the steady state minimax controller

$$u_k = \mu(x_k) = -x_k, \quad k = 1, 2, \ldots \tag{3.40}$$

For the auxiliary game, the pure-feedback saddle-point policy for Player 2, as $\gamma \downarrow \sqrt{2}$, is

$$w_k = \frac{1}{2}x_k, \quad k = 1, 2, \ldots \tag{3.41a}$$

which, together with (3.40), leads to the equilibrium state trajectory (for the game):

$$x_{k+1}^* = (1/2)x_k^*. \tag{3.41b}$$

We can now construct the mixed saddle-point policy for the maximizer in the disturbance attenuation problem, by using the earlier procedure. Using again a normalized (to unity) energy bound, we arrive at (from (3.41a) and (3.41b)):

$$w_k^* = \xi/2^k, \qquad k = 1, 2, \ldots$$

as the worst-case disturbance input.

3.3.3 A least favorable distribution for the disturbance

We are now in a position to obtain a worst-case (*least favorable*) probability distribution for the disturbance in the disturbance attenuation problem of this section. Consider the performance index given by

$$F(\mu, w) := \begin{cases} \|T_\mu(w)\|^2/\|w\|^2, & w_{[1,K]} \neq 0 \\ -\infty, & \text{else} \end{cases} \tag{3.42}$$

Then, if $w^* = \{w_k^*, k \leq K\}$ is a *least favorable* random sequence, we should have

$$\min_{\mu \in \mathcal{M}} E\{F(\mu, w^*)\} = E\{F(\mu_{\hat{\gamma}}^*, w^*)\} = \hat{\gamma}^2 \tag{3.43}$$

where $E\{\cdot\}$ denotes the expectation operator under the probability measure of the least favorable random sequence, and $\mu_{\hat{\gamma}}^*$ is the optimum controller introduced in Theorem 3.5. In words, if the least favorable distribution were made known to the control designer (as *a priori* information), the optimal controller (that now minimizes the expected value of F) would still be the one given in Theorem 3.5. The following Proposition provides such a least favorable sequence.

Discrete-Time Problem with Perfect State Measurements 51

Proposition 3.1. *Let N be a nonnegative definite matrix defined by*

$$N := M_2 + \hat{\gamma}^{-2} \sum_{k=2}^{K} \Phi_{k,2}^{*'} A_k' {\Lambda_k'}^{-1} M_{k+1}' D_k D_k' M_{k+1} \Lambda_k^{-1} A_k \Phi_{k,2}^* \quad (3.44a)$$

where $\Phi_{k,2}^*$ is defined by (using (3.10))

$$x_k^* = \Phi_{k+1}^* x_{k-1}^* =: \Phi_{k,2}^* x_2^* \quad (3.44b)$$

with γ taken to be equal the value $\hat{\gamma}$. Let D_1 have nonzero rank, and η be an eigenvector corresponding to a maximum eigenvalue of the matrix $D_1' N D_1$, where M_2 is generated by (3.3a) with $\gamma = \hat{\gamma}$. Let ξ be a random vector taking the values η and $-\eta$ with equal probability $\frac{1}{2}$. Furthermore, let the random vector sequence $\{\xi_k, k \geq K\}$ be generated by

$$\xi_{k+1} = \Lambda_k^{-1} A_k \xi_k, \quad \xi_2 = D_1 \xi \quad (3.45)$$

where again we take $\gamma = \hat{\gamma}$. Then, a least favorable random sequence for the disturbance is given by

$$w_k^* = \hat{\gamma}^{-2} D_k' M_{k+1} \Lambda_k^{-1} A_k \xi_k, \quad k = 2, \ldots, K; \qquad w_1^* = \xi \quad (3.46)$$

Proof. Even though $\{w_k^*, k \leq K\}$ is a random sequence, it is highly correlated across stages, so that with $u_1 = 0$ we have $x_2 = \xi_2$, which makes the entire future values of ξ_k known to the minimizer. In view of this, the optimality of μ_k^*, $k \geq 2$, follows from Theorem 3.2 with initial state x_2, and by taking the maximizer's policy to be open-loop. Let $n^* := \sum_{k=1}^{K} |w_k^*|^2$, which is independent of the sample path. Then,

$$E\{F(\{\mu_k^*\}_{k=2}^{K}, u_1; \{w_k^*\}_{k=1}^{K})\} = \frac{1}{n^*} E\{x_2' N x_2 + u_1' u_1\}$$

$$= \frac{1}{n^*} E\{|B_1 u_1 + D_1 w_1^*|_N^2 + u_1' u_1\} = \frac{1}{n^*} E\{|D_1 w_1^*|_N^2 + u_1'(B_1' N B_1 + I)u_1\}$$

where the last step follows since w_1^* has a symmetric two-point distribution. Clearly, the unique minimizing solution is $u_1^* = 0$, which completes the proof as far as the optimality of the controller goes. The fact that any one of the two sample paths of $\{w_k^*, k \leq K\}$ maximizes $F(\mu_\gamma^*, w_{[1,K]})$ follows from the saddle-point property of the open-loop version of (3.9) (see Remark 3.2) together with the observation that under a fixed energy bound on the disturbance, w_1^* should maximize the quantity $|D_1 w_1|_N$. Clearly, if the matrix $D_1' N D_1$ has maximum eigenvalues of multiplicity greater than one, the least favorable random sequence may not be unique (even under the hard energy bound, as in the first expression of (1.2b)). ◇

3.3.4 Optimum controller under the 1-step delay information pattern

The counterpart of Theorem 3.5 can be obtained under the 1-step delay information pattern for the controller, by using this time Theorem 3.4. Toward this end, let γ^* be the optimum attenuation level for the perfect-state information case (à la Theorem 3.2), and let Γ_{1D} be the set of all $\gamma \geq \gamma^*$, which further satisfy (3.20a). Let

$$\gamma^\circ := \inf\{\gamma : \gamma \in \Gamma_{1D}\}. \tag{3.47}$$

Then, following a reasoning similar to the one that led to Theorem 3.5, we arrive at the following result for the disturbance attenuation problem with 1-step delayed state measurements.

Theorem 3.6. *Let $\gamma^\circ \geq 0$ be the scalar defined above, and assume that $\Lambda_k(\gamma^\circ)$ is nonsingular for all $k \leq K$.[7] Let $\mu_{\gamma^\circ}^\circ$ be the controller*

$$\mu_k^\circ(\xi_k) = -B_k^T M_{k+1}(\gamma^\circ)\Lambda_k(\gamma^\circ)^{-1} A_k \xi_k, \quad k \in [1, K], \tag{3.48a}$$

$$\xi_{k+1} = A_k x_k + (\Lambda_k(\gamma^\circ)^{-1} - I)A_k \xi_k, \quad \xi_1 = 0. \tag{3.48b}$$

---

[7]This condition holds when $\gamma^\circ > \gamma^*$, which (intuitively) is generally the case, due to loss of information to the controller under delayed state information.

Discrete-Time Problem with Perfect State Measurements

Then, for the discrete-time disturbance attenuation problem with one-step delayed state information,

$$\inf_{\mu \in \mathcal{M}_{\text{CLD}}} \ll T_\mu \gg = \ll T_{\mu^\circ_{\gamma^\circ}} \gg = \gamma^\circ. \tag{3.49}$$

◇

Remark 3.4. If the controller were allowed to depend only on x_{k-1} at stage k, and not also on $x_{[1,k-2]}$, then the only representation of the feedback controller (3.8) in this class would be

$$\begin{aligned}\mu_k(x_{k-1}) &= -B'_k M_{k+1} \Lambda_k^{-1} A_k \Lambda_{k-1}^{-1} A_{k-1} x_{k-1}, \quad k \geq 2 \\ &= B'_1 M_2 \Lambda_1^{-1} A_1 x_1, \quad k = 1 \end{aligned} \tag{3.50}$$

which however requires more stringent conditions on γ than the controller (3.48a). If the resulting performance level is acceptable, however, the advantage here is that the controller is static, while (3.48a) is dynamic (of the same order as that of the state).   ◇

Illustrative Example (continued)

Continuing with the illustrative example of Section 3.3.2, we now study the degradation in performance due to a (one step) delay in the use of the state information, á la Theorem 3.6. Taking $K = 4$, we compute the closed-loop performance bound (from Theorem 3.2) as

$$\gamma^* = 1.4079.$$

For the one-step delay information case, the corresponding performance (as given by (3.47)) is

$$\gamma^\circ = 2.2083,$$

whereas the performance under the simpler (no-memory) controller (3.50) is

$$\gamma^{nm} = 2.5135.$$

Two other possible constructions for the compensator (3.48b) are

(i)

$$\xi_{k+1} = \left(I + \frac{1}{\gamma^2} D_k D'_k M_{k+1} \Lambda_k^{-1}\right) A_k x_k - B_k B'_k M_{k+1} \Lambda_k^{-1} A_k \xi_k, \quad \xi_1 = 0,$$

where we have substituted in the state equation the saddle-point feedback policy (3.8) for the disturbance, as a function of x_k (and not ξ_k); and

(ii)

$$\xi_{k+1} = \left(I - B_k B'_k M_{k+1} \Lambda_k^{-1}\right) A_k \xi_k + \frac{1}{\gamma^2} D_k D'_k M_{k+1} \Lambda_k^{-1} A_k x_k, \quad \xi_1 = 0,$$

where in the state equation we have replaced x_k and u_k by their saddle-point values in terms of ξ_k, but still kept w_k as a function of x_k, as in (i).

The controllers obtained under these two constructions are different representations of the feedback saddle-point policy (3.48a), on the saddle-point trajectory, and hence are themselves also (candidate) saddle-point controllers. However, they should require more stringent conditions on γ (than γ°) in view of Theorem 3.6. Indeed, we find that the best attainable performances in the two cases above are

$$\gamma_{(i)} = 2.2160; \quad \gamma_{(ii)} = 2.7929$$

both of which are inferior to the performance level γ°.

3.4 The Infinite-Horizon Case

The very last result of the *illustrative example* of Section 3.3.2 (i.e., the existence of a unique steady-state minimax controller, obtained as the limit of finite-horizon minimax controllers) prompts the question of whether a characterization of a stationary minimax controller can be obtained for the time-invariant version of the general disturbance attenuation problem of Section 2. Said another way, the question is whether Theorem 3.5 admits a limiting case, as $K \to \infty$, when all the system matrices are time-invariant.[8]

---
[8]Henceforth, in this section, we take all the system matrices, in (3.1) and (3.2a), as constant matrices.

A further question is whether the resulting stationary controller is stabilizing. And a third question is whether the stabilizing stationary controller obtained as the limit of the finite-horizon optimal controller is indeed optimal for the infinite-horizon disturbance attenuation problem. The answers to all three questions are in the affirmative, under appropriate conditions on the system matrices, as to be established below.

For future reference, let us first write down the steady-state version of the feedback saddle-point solution of the soft-constrained dynamic game covered by Theorem 3.2, along with the steady-state versions of (3.4a), (3.4b) and (3.7). In all cases the limiting values (as $K \to \infty$) are designated by an "overbar."

Saddle-point controllers:

$$\bar{u}_k = \bar{\mu}(x_k) = -B'\bar{M}(\gamma)\bar{\Lambda}^{-1}(\gamma)Ax_k, \quad k \geq 1 \tag{3.51a}$$

$$\bar{w}_k = \bar{\nu}(x_k) = \gamma^{-2}D'\bar{M}(\gamma)\bar{\Lambda}^{-1}(\gamma)Ax_k, \quad k \geq 1 \tag{3.51b}$$

$$\bar{\Lambda}(\gamma) := I + \left(BB' - \gamma^{-2}DD'\right)\bar{M}(\gamma) \tag{3.52a}$$

and $\bar{M}(\gamma)$ satisfies the (generalized) algebraic Riccati equation (ARE)

$$\bar{M}(\gamma) = Q + A'\bar{M}(\gamma)\bar{\Lambda}^{-1}(\gamma)A. \tag{3.52b}$$

Existence condition (as the counterpart of (3.7)):

$$\gamma^2 I - D'\bar{M}(\gamma)D > 0. \tag{3.53}$$

Toward validating the above, we now first establish the monotonicity of the sequence generated by (3.4).

Lemma 3.3. *Given an integer K, let γ be chosen such that condition (3.7) of Theorem 3.2 is satisfied for all $k \leq K$. Let M_k, $k = K, K-1, ..., 1$, be the sequence of nonnegative definite matrices generated by iteration (3.4a). Then,*

$$M_k \geq M_{k+1} \quad \text{for all } k \leq K$$

(i.e., $M_k - M_{k+1}$ is a nonnegative definite matrix).

Proof. The proof is similar to that of monotonicity of the solution of Riccati equation in linear-quadratic control, but now we use the saddle-point value function (of the soft-constrained game) rather than the dynamic programming value function. Toward this end, we first write down the *Isaacs'* equation (2.12), associated with the linear-quadratic "feedback" game of Section 3.2.2:

$$V_k(x) = \min_u \{V_{k+1}(Ax + Bu + Dw) + |x|_Q^2 + |u|^2 - \gamma^2|w|^2\}$$

$$V_{K+1}(x_{K+1}) = |x_{K+1}|_Q^2$$

Here $V_k(x)$ is the saddle-point value of the dynamic game of Section 3.2.2, with only $K-k+1$ stages (i.e., starting at stage k with state x and running through K). We know from Theorem 3.2 that[9] $V_k(x) = |x|_{M_k}^2$, and

$$|x|_{M_k}^2 = \min_u \max_w \{|Ax + Bu + Dw|_{M_{k+1}}^2 + |x|_Q^2 + |u|^2 - \gamma^2|w|^2\} \quad (*)$$

for every $x \in \mathbb{R}^n$. Now, since for two functionals g and f,

$$g(u,w) \geq f(u,w) \Rightarrow \inf_u \sup_w g(u,w) \geq \inf_u \sup_w f(u,w),$$

it follows from (∗) that the matrix inequality $M_{k+1} \geq M_{k+2}$ implies the inequality $M_k \geq M_{k+1}$. Hence, the proof of the lemma will be completed (by induction) if we can show that $M_K \geq M_{K+1} = Q$. Since we know that under condition (3.5) with $k = K$,

$$Q\Lambda_K^{-1} = Q[I + (BB' - \gamma^{-2}DD')Q]^{-1} \geq 0,$$

it follows that

$$M_K = Q + A'Q\Lambda_K^{-1}A \geq Q = M_{K+1}$$

thus completing the proof. ◇

The next lemma provides a set of conditions under which the sequence $\{M_{k,K+1}\}_{k=K+1}^1$ is bounded, for every $K > 0$. Here we use a double index

---
[9] This fact has already been used in the proof of Theorem 3.2.

Discrete-Time Problem with Perfect State Measurements

on M to indicate explicitly the dependence of the sequence on the terminal (starting) time point K. Of course, for the time-invariant problem, the elements of the sequence will depend only on the difference, $K - k$, and not on k and K separately.

Lemma 3.4. *Let $\gamma > 0$ be fixed, and $\bar{M}$ be a positive definite solution of the ARE (3.52b), which also satisfies the condition (3.53). Then, for every $K > 0$,*

$$\bar{M} - M_{k,K+1} \geq 0, \quad \text{for all } k \leq K + 1 \tag{3.54}$$

where $M_{k,K}$ is generated by (3.4a).

Proof. First note that (3.53) implies nonsingularity of $\bar{\Lambda}$, and since by hypothesis $\bar{M} > 0$, it follows that $\bar{M}\bar{\Lambda}_{-1} > 0$. If this property is used in (3.52b), we immediately have $\bar{M} \geq Q$, which proves (3.54) for $k = K+1$. We now show that the validity of (3.54) for $k+1$ implies its validity for k. Toward this end let us first assume that $M_{k+1,K+1} > 0$, and note from (3.4a) and (3.52b), that

$$\bar{M} - M_{k,K+1} = A' \left(\bar{M}\bar{\Lambda}^{-1} - M_{k+1,K+1}\Lambda_{k,K+1}^{-1} \right) A$$

$$= A' \left(\left[\bar{M}^{-1} + BB' - \frac{1}{\gamma^2}DD' \right]^{-1} \right.$$

$$\left. - \left[M_{k+1,K+1}^{-1} + BB' - \frac{1}{\gamma^2}DD' \right]^{-1} \right) A$$

which is nonnegative definite since $\bar{M} \geq M_{k+1,K+1}$ by the hypothesis of the inductive argument. By the continuous dependence of the eigenvalues of a matrix on its elements the result holds also for $M_{k+1,K+1} \geq 0$, since we can choose a matrix $N(\epsilon)$, and a sufficiently small positive parameter ϵ_0 such that $0 < N(\epsilon) \leq \bar{M}$, $0 < \epsilon < \epsilon_0$, and $N(0) = M_{k+1,K+1}$. This completes the proof of the lemma. ◊

The next lemma says that any nonnegative definite solution of (3.52b) has to be positive definite, if we take the pair (A, H) to be observable, where $H'H = Q$ (this requirement henceforth referred to as "$(A, Q^{1/2})$ being observable"). The proof of this result is similar to that of the standard ARE which arises in linear regulatory theory [56]; it is given below for the sake of completeness.

Lemma 3.5. *Let $(A, Q^{1/2})$ be observable. Then, if there exists a nonnegative definite solution to the ARE (3.52b), satisfying (3.53), it is positive definite.*

Proof. Let $H'H := Q$. Assume that, to the contrary, ARE has a nonnegative definite solution with at least one zero eigenvalue. Let x be an eigenvector corresponding to this eigenvalue. Then,

$$x'\bar{M}x = x'H'Hx + x'A'\bar{M}\bar{\Lambda}^{-1}Ax = 0$$

$$\Rightarrow \quad Hx = 0 \text{ and } \bar{M}^{\frac{1}{2}}Ax = 0$$

where $\bar{M}^{\frac{1}{2}}$ is the unique nonnegative definite square-root of $\bar{M}$, and the last result follows because both terms on the right are nonnegative definite quadratic terms, and $\bar{M}\bar{\Lambda}^{-1}$ can equivalently be written as

$$\bar{M}^{\frac{1}{2}}(I + \bar{M}^{\frac{1}{2}}(BB' - \gamma^{-2}DD')\bar{M}^{\frac{1}{2}})^{-1}\bar{M}^{\frac{1}{2}}.$$

Next, multiply the ARE from left and right by $x'A'$ and Ax, respectively, to arrive at

$$x'A'\bar{M}Ax = x'A'H'HAx + x'(A')^2\bar{M}\bar{\Lambda}^{-1}A^2x.$$

The left-hand side is *zero* by the earlier result, and hence

$$HAx = 0 \quad \text{and} \quad \bar{M}^{\frac{1}{2}}A^2x = 0.$$

Now multiply the ARE from left and right by $x'(A^2)'$ and A^2x, respectively, and continue in the manner above to finally arrive at the relation

$$x'\left(H', A'H', (A^2)'H', \ldots, (A^{n-1})'H'\right) = 0$$

which holds, under the observability assumption, only if $x = 0$. Hence $\bar{M} > 0$. ◇

An important consequence of the above result, also in view of Lemmas 3.3 and 3.4, is that if there exist multiple positive-definite solutions to the ARE (3.52b), satisfying (3.53), there is a minimal such solution (minimal in the sense of matrix partial ordering), say $\bar{M}^+$, and that

$$\lim_{K \to \infty} M_{k,K+1} = \bar{M}^+ > 0,$$

whenever (A,C) is an observable pair. Clearly, this minimal solution will also determine the value of the infinite-horizon game, in the sense that

$$\inf_{\mu_{[1,\infty)} \in \mathcal{M}_{\text{CLPS}}^\infty} \sup_{w_{[1,\infty)}} J_\gamma^\infty(\mu_{[1,\infty)}, w_{[1,\infty)}) = x_1' \bar{M}^+(\gamma) x_1$$

$$= \sup_{\nu_{[1,\infty)} \in \mathcal{N}_{\text{CLPS}}^\infty} \inf_{\mu_{[1,\infty)} \in \mathcal{M}_{\text{CLPS}}^\infty} J_\gamma^\infty(\mu_{[1,\infty)}, \nu_{[1,\infty)})$$

(3.55)

where J_γ^∞ is (3.2b) with $K = \infty$.

The following lemma now says that the existence of a positive definite solution to the ARE (3.52b), satisfying (3.53), is not only sufficient, but also necessary for the value of the game to be bounded, again under the observability condition.

Lemma 3.6. *Let* $(A, Q^{1/2})$ *be observable. Then, if the ARE (3.52b) does not admit a positive definite solution satisfying (3.53), the upper value of the game* $\{J_\gamma^\infty; \mathcal{M}_{\text{CLPS}}^\infty), \mathcal{H}_w^\infty\}$ *is unbounded.*

Proof. Since the limit point of the monotonic sequence of nonnegative definite matrices $\{M_{k,K+1}\}$ has to constitute a solution to (3.52b), nonexistence of a positive definite solution to (3.52b) satisfying (3.53), (which also implies nonexistence of a nonnegative definite solution, in view of Lemma 3.5) implies that for each fixed k, the sequence $\{M_{k,K+1}\}_{K>0}$ is unbounded. This means that given any (sufficiently large) $\alpha > 0$, there exists a $K > 0$, and an initial state $x_1 \in \mathbb{R}^n$, such that the value of the K-stage game exceeds $\alpha |x_1|^2$. Now choose $w_k = 0$ for $k > K$. Then,

$$\inf_\mu \sup_w J_\gamma^\infty(\mu, w) \geq \inf_\mu \left\{ \sum_{k=K+1}^\infty \{|x_{k+1}|_Q^2 + |u_k|^2\} \right.$$

$$\left. + \sum_{k=1}^K \{|x_{k+1}|_Q^2 + |u_k|^2 - |\nu_k^*(x_k)|^2\} + |x_1|_Q^2 \right\}$$

$$\geq x_1' M_{1,K+1} x_1 > \alpha |x_1|^2$$

which shows that the upper value can be made arbitrarily large. In the above, $\nu_{[1,K]}^*$ in the first inequality is the feedback saddle-point controller for Player 2 in the K-stage game, and the second inequality follows because the summation from $K+1$ to ∞ is nonnegative and hence the quantity is bounded from below by the value of the K-stage game. ◇

Lemmas 3.3–3.6 can now be used to arrive at the following counterpart of Theorem 3.2 in the infinite-horizon case.

Theorem 3.7. *Consider the infinite-horizon discrete-time linear-quadratic soft-constrained dynamic game, with $\gamma > 0$ fixed and $(A, Q^{1/2})$ constituting an observable pair. Then,*

(i) *The game has equal upper and lower values if, and only if, the ARE (3.52b) admits a positive definite solution satisfying (3.53).*

(ii) *If the ARE admits a positive definite solution, satisfying (3.53), then it admits a minimal such solution, to be denoted $\bar{M}^+(\gamma)$. Then, the finite value of the game is (3.55).*

(iii) *The upper (minimax) value of the game is finite if, and only if, the upper and lower values are equal.*

(iv) *If $\bar{M}^+(\gamma) > 0$ exists, as given above, the controller $\bar{\mu}_{[1,\infty)}$ given by (3.51a), with $\bar{M}(\gamma)$ replaced by $\bar{M}^+(\gamma)$, attains the finite upper value,* *in the sense that*

$$\sup_{w_{[1,\infty)} \in \mathcal{H}_w^\infty} J_\gamma^\infty(\bar{\mu}_{[1,\infty)}, w_{[1,\infty)}) = x_1' \bar{M}^+(\gamma) x_1 , \qquad (3.56)$$

and the maximizing feedback solution above is given by (5.1b), again with $\bar{M}$ replaced by $\bar{M}^+$.

(v) Whenever the upper value is bounded, the feedback matrix

$$F := (I - BB'\bar{M}^+(\gamma)\bar{\Lambda}^+(\gamma)^{-1})A \qquad (3.57a)$$

is Hurwitz, that is it has all its eigenvalues inside the unit circle. This implies that the linear system

$$x_{k+1} = Fx_k + Dw_k \qquad (3.57b)$$

is bounded input–bounded output stable.

Proof. Parts (i)–(iii) follow from the sequence of Lemmas 3.3–3.6, as also discussed prior to the statement of the theorem. To prove part (iv), we first note that the optimization problem in (3.56) is the maximization of

$$\sum_{k=1}^\infty \left\{ |x_{k+1}|_Q^2 + |B'\bar{M}^+\bar{\Lambda}^{+^{-1}} Ax_k|^2 - \gamma^2 |w_k|^2 \right\} + |x_1|_Q^2$$

over $w_{[1,\infty)}$, subject to the state equation constraint (3.57b). First consider the truncated (K-stage) version:

$$\max_{w_{[1,K]}} \sum_{k=1}^K \left\{ |x_{k+1}|_Q^2 + |B'\bar{M}^+\bar{\Lambda}^{+^{-1}} Ax_k|^2 - \gamma^2 |w_k|^2 \right\} + |x_1|_Q^2$$

$$\leq \max_{w_{[1,K]}} \left\{ |x_{K+1}|_{\bar{M}^+}^2 + \sum_{k=1}^K |x_k|_{\tilde{Q}_k}^2 - \gamma^2 |w_k|^2 \right\} + |x_1|_Q^2$$

where

$$\tilde{Q}_k := \begin{cases} Q + A'(\bar{M}^+\bar{\Lambda}^{+^{-1}})'BB'\bar{M}^+\bar{\Lambda}^{+-1}A, & k > 1 \\ A'(\bar{M}^+\Lambda^{+^{-1}})'BB'\bar{M}^+\bar{\Lambda}^{+^{-1}}A, & k = 1 \end{cases}$$

with the inequality following because $\bar{M}^+ \geq Q$. Adopting a dynamic programming approach to solve this problem, we have, for the last stage, after some manipulations:

$$\max_w \left\{ |Fx_K + Dw|^2_{\bar{M}^+} + |x_K|^2_{\bar{Q}_K} - \gamma^2 |w|^2 \right\} = |x_K|^2_{\bar{M}^+},$$

which is uniquely attained by

$$\begin{aligned} w^* &= (\gamma^2 I - D'\bar{M}^+ D)^{-1} D' \bar{M} F x_K \\ &\equiv (1/\gamma^2) D' \bar{M}^+ \bar{\Lambda}^{+^{-1}} A x_K \end{aligned} \quad (*)$$

under the strict concavity condition

$$\gamma^2 I - D'\bar{M}^+ D > 0,$$

which is (3.53) and is satisfied by hypothesis. Hence recursively we solve identical optimization problems at each stage, leading to

$$\max_{w_{[1,K]}} \sum_{k=1}^{K} \left\{ |x_{k+1}|^2_Q + |B'\bar{M}^+ \bar{\Lambda}^{+^{-1}} A x_k|^2 - \gamma^2 |w_k|^2 \right\} \leq x_1^T \bar{M}^+ x_1$$

where the bound is independent of K. Since we already know from part (ii) that this bound is the value of the game, it readily follows that the control $\bar{\mu}_{[1,\infty)}$ attains it, and the steady state controller $(*)$ maximizes $J_\gamma^\infty(\bar{\mu}_{[1,\infty)}, w_{[1,\infty)})$.

Now, finally to prove part (v), we use an argument similar to that used in linear regulator theory. Toward this end, we first note that boundedness of the upper value implies, with $w_k \equiv 0$, $k \geq 1$, that

$$|x_k|^2_Q + |\bar{\mu}_k(x_k)|^2 \to 0 \quad \text{as} \quad k \to \infty$$

$$\iff \quad \begin{aligned} Hx_k &\to 0 \quad \text{and} \quad Sx_k \to 0, \\ S &:= B'\bar{M}^+\bar{\Lambda}^{+^{-1}} A; \quad H'H := Q \\ Hx_{k+1} &\to 0 \Rightarrow H(A+BS)x_k \to 0 \\ &\Rightarrow HAx_k \to 0 \\ &\cdots \quad \cdots \quad \cdots \\ Hx_{k+n-1} &\to 0 \Rightarrow \quad \cdots \quad \Rightarrow HA^{(n-1)}x_k \to 0. \end{aligned}$$

But $HA^i x_k \to 0$, $i = 0, \ldots, n-1$, implies by *observability* that $x_k \to 0$, and hence F is stable since x_k is generated here by (3.57b), with $w_k \equiv 0$. ◇

Note that the theorem above does *not* claim that the policies (3.51a) and (3.51b) with $\bar{M} = \bar{M}^+$ are in saddle-point equilibrium. Part (iv) only says that with $\mu_{[1,\infty)}$ fixed at $\bar{\mu}_{[1,\infty)}$, $\bar{\nu}_{[1,\infty)}$ maximizes $J_\gamma^\infty(\bar{\mu}_{[1,\infty)}, w_{[1,\infty)})$, which is only one side (left-hand side) of the saddle-point inequality (2.11a). This, however, is sufficient for the disturbance attenuation problem under consideration, since our interest lies only in the upper value of the soft-constrained game. In view of this, the solution to the infinite-horizon version of the disturbance attenuation problem follows from Theorem 3.7 above, by following the line of reasoning that led from Theorem 3.2 to Theorem 3.5. The result is given below as Theorem 3.8:

Theorem 3.8. *For the time-invariant infinite-horizon disturbance attenuation problem of this section, assume that (A, B) is stabilizable and $(A, Q^{1/2})$ is observable. Then:*

(i) *There exits a scalar $\hat{\gamma} > 0$ such that for every $\gamma > \hat{\gamma}$, the nonlinear algebraic equation (ARE) (3.52a) admits a minimal positive definite solution $\bar{M}^+(\gamma)$ in the class of nonnegative definite matrices which further satisfy the condition (3.53).*

(ii) *Let $\hat{\Gamma}$ be the set of nonnegative $\hat{\gamma}$'s satisfying the condition of (i) above. Let $\hat{\gamma}_\infty^* := \inf\{\hat{\gamma} > 0 : \hat{\gamma} \in \hat{\Gamma}\}$. Then, the stationary feedback controller* [10]

$$u_k = \bar{\mu}^*(x_k) = -B'\bar{M}^+(\hat{\gamma}_\infty^*)\bar{\Lambda}^+(\hat{\gamma}_\infty^*)^{-1}Ax_k, \quad k = 1, 2, \ldots \quad (3.58)$$

solves the infinite-horizon disturbance attenuation problem, achieving an attenuation level of $\gamma_\infty^* = \hat{\gamma}_\infty^*$, i.e.,

$$\gamma_\infty^* := \inf_{\mu \in \mathcal{M}_{\text{CLPS}}^\infty} \ll T_\mu \gg = \ll T_{\bar{\mu}^*} \gg = \hat{\gamma}_\infty^*$$

Furthermore, the controller (3.58) leads to a bounded input–bounded output stable system.

---
[10] Here the gain coefficient of the controller remains bounded as $\gamma \downarrow \gamma_\infty^*$ by (the infinite-horizon version of) Lemma 3.2.

Proof. First, we know from the relationship between the disturbance attenuation problem and the related soft-constraint game that γ_∞^* is the optimum attenuation level for the former if, and only if, the game with the objective function J_γ^∞ has a bounded upper value for all $\gamma > \gamma_\infty^*$, and the upper value is unbounded for for $\gamma < \gamma_\infty^*$. This, on the other hand, holds (in view of Theorem 3.7 and under the observability condition) if, and only if, (3.52a) admits a positive definite solution, satisfying (3.53), which is also the minimal such solution. Hence, all we need to show is that there exists some finite $\hat{\gamma}$, so that for all $\gamma > \hat{\gamma}$ the upper value is bounded. It is at this point that the first condition (on stabilizability) of the Theorem is used, which ensures that the upper value is bounded for sufficiently large $\hat{\gamma}$, since as $\gamma \uparrow \infty$ the LQ game reduces to the standard linear regulator problem[56]. A continuity argument proves the result. ◇

Remark 3.5. Returning to the earlier illustrative example of Section 3.3.2, where we had found (through a limiting approach) that $\gamma_\infty^* = \sqrt{2}$, let us now directly use Theorem 3.8. First, solving for $\bar{M}(\gamma)$ from (3.52a), we obtain

$$\bar{M}(\gamma) = \frac{1 \pm \sqrt{(5\gamma^2 - 1)/(\gamma^2 - 1)}}{2}$$

which is nonnegative if, and only if, we choose the positive square root, and (3.53) holds if and only if $\gamma > \sqrt{2}$. This readily leads to $\gamma_\infty^* = \sqrt{2}$ as the attenuation constant. The resulting closed-loop system is $x_{k+1} = w_k$, which is clearly bounded input–bounded output stable. ◇

Remark 3.6. It is possible to obtain the infinite-horizon version of Theorem 3.6 for the time-invariant problem, by essentially using the result of Theorem 3.8. First we invoke the two conditions (on stabilizability and observability) of Theorem 3.8, under which for $\gamma > \gamma_\infty^*$ we have the steady-state controller (from (3.48a)):

$$\mu_\gamma^o(\xi_k) = -B'\bar{M}^+(\gamma)\bar{\Lambda}^+(\gamma)^{-1}A\xi_k \qquad (3.59a)$$

where ξ_k is now generated by

$$\xi_{k+1} = -(I - \bar{\Lambda}^+(\gamma)^{-1})A\xi_k + Ax_k. \tag{3.59b}$$

One then has to bring in additional conditions, for the stability of the delayed linear system[11], so that (3.59b) along with the system

$$x_{k+1} = Ax_k - BB'\bar{M}^+(\gamma)\bar{\Lambda}^+(\gamma)^{-1}A\xi_k + Dw_k$$

is stable, and the solution of (3.20b), $\hat{S}_{k,K+1}$, converges to a well-defined limit as $K \to \infty$ (say, $\bar{S}(\gamma)$). Let

$$\Gamma_\infty^\circ := \left\{ \gamma \geq \gamma_\infty^* : \gamma^2 I - \hat{D}'\bar{S}(\gamma)\hat{D} > 0 \right\}. \tag{3.60a}$$

Then, the minimax attenuation level is

$$\gamma_\infty^\circ := \inf \{\gamma : \gamma \in \Gamma_\infty^\circ\}. \tag{3.60b}$$

◇

3.5 More General Classes of Problems

In this section we discuss some extensions of the results of the previous sections in two directions, to accommodate *general cost functions of the type* (1.3a), and *nonzero initial states*.

3.5.1 More general plants and cost functions

We return here to the original class of disturbance attenuation problems formulated by (1.6a), (1.6b) and (1.7a), and consider two cases. The first is the direct extension of the formulation (3.1)-(3.2), where the cost function has an additional cross term between the state and the control, namely

$$L(u,w) = |x_{K+1}|^2_{Q_f} + \sum_{k=1}^{K} |z_k|^2 \tag{3.61}$$

---

[11] One such sufficient condition is $Q > 0$, under which boundedness of upper value implies input-output stability under controller (3.59a).

$$z_k = H_k x_k + G_k u_k \tag{3.62}$$

where we assume that

$$G'_k G_k =: R_k > 0, \quad k \in [1, K]. \tag{3.63a}$$

Furthermore, by hypothesis,

$$H'_k G_k =: P_k \neq 0. \tag{3.63b}$$

Now, using the linear transformation

$$u_k \rightarrow \bar{u}_k := R_k^{\frac{1}{2}}(u_k + R_k^{-1} P'_k x_k), \tag{3.64}$$

we can rewrite (3.61) as

$$L(u, w) = |x_{K+1}|^2_{Q_f} + \sum_{k=1}^{K} |x_k|^2_{\bar{Q}_k} + |\bar{u}_k|^2 \tag{3.65a}$$

$$\bar{Q}_k := Q_k - P_k R_k^{-1} P'_k. \tag{3.65b}$$

Note that (3.65a) is identical to (3.2a), with the control input now being $\bar{u}_{[1,K]}$. In terms of this input, the system equation (3.1) becomes

$$x_{k+1} = \bar{A}_k x_k + \bar{B}_k \bar{u}_k + D_k w_k, \tag{3.66a}$$

$$\bar{A}_k := A_k - B_k R_k^{-\frac{1}{2}}. \tag{3.66b}$$

Hence, whenever $\bar{Q}_k \geq 0$, $k \in [1, K]$, this problem is no different from the one solved in Section 3.3.1, and Theorem 3.5 equally applies here, with A_k, B_k, Q_k replaced by $\bar{A}_k$, $\bar{B}_k$ and $\bar{Q}_k$, respectively, Analogously, the counterpart of Theorem 3.8 also holds here, provided that the pair $(\bar{A}, \bar{Q}^{\frac{1}{2}})$ is observable.

As a second extension, consider again the general formulation of (1.6a)-(1.7a), but with (1.7a) now written as

$$L(u, w) = |\zeta_{K+1}|^2_{Q_f} + \sum_{k=1}^{K} \{|\zeta_k|^2_{Q_k} + |u_k|^2\}. \tag{3.67}$$

Note that this corresponds to (1.6b) with

$$H'_k H_k =: Q_k, \quad H'_k G_k = 0, \quad G'_k G_k = I. \quad (3.68)$$

Here ζ_k can be viewed as part of the "controlled output", to which there is a direct link from both the disturbance and the control in a *strictly causal* sense. Even though this seems, at the outset, to be a much more general class of problems than the ones considered heretofore, we will show below that the results of Theorems 3.5 and 3.8 are directly applicable here as well. This is done by first expanding and then appropriately contracting the state space.[12] Toward this end, let us view $\zeta_k, k \in [1, K]$, as an additional state variable, and introduce the state dynamics

$$t_{k+1} = \widetilde{A}_k t_k + \widetilde{B}_k u_k + \widetilde{D}_k w_k \quad (3.69a)$$

where

$$t_k := (\zeta'_k, x'_k)' \quad (3.69b)$$

$$\widetilde{A}_k := \begin{pmatrix} 0 & \widehat{H}_{k+1} A_k \\ 0 & A_k \end{pmatrix}, \quad \widetilde{B}_k := \begin{pmatrix} \widehat{H}_{k+1} B_k + \widehat{G}_k \\ B_k \end{pmatrix},$$

$$\widetilde{D}_k := \begin{pmatrix} \widehat{H}_{k+1} D_k + \widehat{F}_k \\ D_k \end{pmatrix}. \quad (3.69c)$$

Then, the performance index (1.7a) can be written as

$$J' = \sum_{k=1}^{K} \{|t_{k+1}|^2_{\widetilde{Q}_{k+1}} + |u_k|^2\} \quad (3.70)$$

where $\widetilde{Q}_k :=$ block diag $(Q_k, 0)$. Thus, the problem is cast in the framework of the "state-output" problem of the previous sections, making the developed theory applicable here as well, provided that we replace all the system matrices by the new ones defined above. Note that since

$$\widetilde{A}_k t_k = \begin{pmatrix} H_{k+1} \\ I \end{pmatrix} A_k x_k, \quad (3.71)$$

---
[12] For a related transformation in the time-invariant continuous-time case, see [74] which uses a loop-shifting method.

the unique FB saddle-point policies depend only on x_k (and not also on ζ_k), thus making the solution compatible with the underlying information pattern. Perhaps another useful observation here is that even though the main recursion (3.3) now appears (at the outset) to be of dimension $2n \times 2n$, $\widetilde{M}_k$ is in fact block diagonal, with the first block being equal to Q_k. Hence, one needs to iterate only on the second $(n \times n)$ block of $\widetilde{M}_k$, thus making the recursion only of dimension $n \times n$, as in (3.4a).

3.5.2 Nonzero Initial State

If the initial state x_1 is a known nonzero quantity, the equivalence in (1.2b) does not hold, even if the controllers are restricted to the linear class. In this case the first of these (with the hard bound on the norm of the disturbance) is more meaningful. Some recent work ([34]) has shown that its solution satisfies a threshold property, with the threshold determined by some norm of x_1 and by the given norm bound on w. We will not discuss this class of problems here, as its coverage would divert us from our main objective in this book.

If, on the other hand, the initial state is an unknown quantity, one approach would be to consider it also as part of the disturbance, and to obtain the minimax controller under "worst" possible values of x_1. We then replace the norm of $w_{[1,K]}$ in (3.26b) by the quantity

$$|x_1|^2_{Q_0} + \sum_{k=1}^{K} |w_k|^2 \qquad (3.72)$$

which really amounts to adding an extra stage to the problem, with

$$w_0 := Q_0^{1/2} x_1 \implies x_1 = x_0 + w_0, \ x_0 = 0.$$

Hence, we now have a design problem that has $K+1$ stages instead of K, again with *zero* initial state, thus making the earlier results directly applicable. Note that in this reformulation we could take the disturbance at stage *zero* to have the same dimension as the state, because for the finite-horizon problem no restrictions were imposed on the dimensions of the disturbance across stages.

An alternative way of accommodating this uncertainty in x_1 in the solution process would be to carry out the derivation (in the soft-constrained game) as though x_1 was a known nonzero quantity (as we have already done; see Theorems 3.2 and 3.7), and then do further maximization over $x_1 \in \mathbb{R}^n$. Since the upper values of the soft-constrained finite- and infinite-horizon games with CLPS information pattern are quadratic functions of x_1 (see (3.11) and (3.56), respectively), it readily follows that the maximizing x_1 will be *zero*, provided that the upper value is bounded. For the boundedness, we will need, in addition to the conditions of Theorem 3.5 or Theorem 3.8, the respective conditions:

$$Q_0 - M_1(\gamma) \geq 0, \qquad (3.73a)$$

or

$$Q_0 - \bar{M}^+(\gamma) \geq 0, \qquad (3.73b)$$

which may (or may not) lead to an increase in the minimax attenuation level, depending on the value of the weighting matrix Q_0. Now if the information structure is not CLPS, the approach above is still valid, since the saddle-point value of the soft-constrained game does not depend on the particular information pattern, provided that it exists.[13] Hence, if x_1 is an unknown quantity, (3.65a) or (3.65b) (depending on whether we have finite or infinite horizon) constitutes the additional condition to be satisfied, under all information patterns.

3.6 Extensions to Nonlinear Systems under Nonquadratic Performance Indices

The general approach adopted in this chapter can be also be used for systems with a nonlinear state equation and/or a nonquadratic performance

---
[13] This follows from Theorem 2.5.

index. Consider the nonlinear system described by (2.7a), with performance index (2.8), and initial state $x_1 = 0$. Choose, as the associated soft-constrained dynamic game, the one with dynamics (2.7a) and cost function

$$L_\gamma(u, w) = L(u, w) - \gamma^2 \|w\|^2, \qquad (3.74)$$

where L is defined by (2.7a). As discussed in Section 2.2, a feedback saddle-point solution for this soft-constrained game is given by the Isaacs' equation (2.12), where in this case we replace g by g^γ, defined by

$$g_k^\gamma(x_{k+1}, u_k, w_k, x_k) := g_k(x_{k+1}, u_k, w_k, x_k) - \gamma^2 |w_k|^2. \qquad (3.75)$$

Let the value function generated by (2.12) (with g replaced by g^γ) be $V_k^\gamma(x)$, $k \in [1, K]$, and the corresponding minimizing feedback control (assuming that it exists) be

$$u_k^* = \mu_k^\gamma(x_k), \quad k \in [1, K]. \qquad (3.76)$$

Finally, introduce the following counterpart of the threshold value $\hat{\gamma}$ introduced in Section 3.3.1:

$$\hat{\gamma}^{\text{NCL}} := \inf\{\gamma \in \Gamma^{\text{NCL}}\} \qquad (3.77a)$$

$$\Gamma^{\text{NCL}} := \{\tilde{\gamma} > 0 : \forall \gamma > \tilde{\gamma}, \ V^\gamma \text{ exists}\}. \qquad (3.77b)$$

We now have the following result on guaranteed disturbance bounds, which follows from Theorem 2.4, and the reasoning that led to Theorem 3.5.

Theorem 3.9. *For the nonlinear/nonquadratic disturbance attenuation problem formulated above, let the value function V^γ satisfy the property:*

$$V_1^\gamma(0) = 0, \quad \forall \gamma > \hat{\gamma}^{\text{NCL}}. \qquad (3.78)$$

Then, for every $\gamma > \hat{\gamma}^{\text{NCL}}$, the feedback controller $\mu^\gamma = \mu_{[1,K]}^\gamma$ ensures a disturbance attenuation bound γ, that is

$$\|T_{\mu^\gamma}(w)\|^2 := J(\mu^\gamma, w) \le \gamma^2 \|w\|^2, \quad \forall \, w \in \mathcal{W}. \qquad \diamond$$

Even though this result is the counterpart of Theorem 3.5 for the nonlinear problem, it is weaker than Theorem 3.5 because there is no claim of optimality of the given threshold level $\hat{\gamma}^{NCL}$. For a stronger result, one has to assume some further structure on the functions f and g.

3.7 Summary of Main Results

This chapter has provided a complete solution to the basic discrete-time disturbance attenuation problem in both finite and infinite horizons, when the controller has access to the state with or without delay. In the latter case the optimum attenuation bound and the corresponding optimum controller are given in Theorems 3.5 and 3.8, for the finite- and infinite-horizon cases, respectively. The optimum controller is a memoryless feedback controller, given in terms of the solution of a discrete-time Riccati equation (3.4a), or an algebraic Riccati equation (3.52b). In the delayed measurement case, the optimum attenuation level and the corresponding optimum controller are given for the finite-horizon case in Theorem 3.6, with extension to infinite horizon discussed in Remark 3.6. An interesting feature of the solution is that the optimal controller uses an n-dimensional compensator, which is given by (3.48b), or (3.59b).

In the chapter we have also discussed extensions of the results obtained for the basic problem to more general classes of problems which involve (i) cross terms in the performance index, (ii) direct delayed dependence of the output on the disturbance and the control input, (iii) nonzero initial states, (iv) nonlinear plants and/or nonquadratic performance indices.

This chapter is based on work first presented in [6], and then expanded in [9] [10]. Some parallel results on the discrete-time disturbance attenuation problem have been obtained in [60], [92], [77], [78].

Chapter 4

Continuous-Time Systems With Perfect State Measurements

4.1 Introduction

In this chapter, we develop the continuous-time counterparts of the results presented in Chapter 3. The disturbance attenuation problem will be the one formulated in Chapter 1, through (1.9)-(1.10), under only perfect state measurements. This will include the closed-loop perfect state (CLPS), sampled-data perfect state (SDPS), and delayed perfect state (DPS) information patterns, which were introduced in Chapter 2 (Section 2.3).

For future reference, let us first reintroduce the essential elements of the continuous-time problem when the controlled output is a simple concatenation of the state and the control :

$$\frac{d}{dt}x(t) =: \dot{x} = A(t)x + B(t)u(t) + D(t)w(t), \quad t \geq 0 \qquad (4.1)$$

$$L(u,w) = |x(t_f)|^2_{Q_f} + \int_0^{t_f} \left\{ |x(t)|^2_{Q(t)} + |u(t)|^2 \right\} dt \qquad (4.2a)$$

$$L_\gamma(u,w) = |x(t_f)|^2_{Q_f} + \int_0^{t_f} \left\{ |x(t)|^2_{Q(t)} + |u(t)|^2 - \gamma^2 |w(t)|^2 \right\} dt \qquad (4.2b)$$

where $Q_f \geq 0$, $Q(t) \geq 0$, $t \in [0, t_f]$, $\gamma > 0$, t_f is the terminal time, all matrices have piecewise continuous entries, and so do u and w. In the above, (4.1) is the differential equation describing the evolution of the state, under the control (u) and disturbance (w) inputs; (4.2a) is the performance index of the original (hard-constrained) disturbance attenuation problem;

and (4.2b) is the performance index of the associated soft-constrained differential game. Note that in terms of the notation of (1.9b),

$$H'(t)G(t) = 0, \quad H'(t)H(t) =: Q(t), \quad G'(t)G(t) = I.$$

(4.1), (4.2a) and (4.2b) are in fact the direct counterparts of (3.1), (3.2a) and (3.2b), respectively, in the continuous time. As in Chapter 3, we will alternatively use the notation $J_\gamma(\mu, w)$, for L_γ, when the control u is generated by a particular policy $\mu \in \mathcal{M}$, compatible with the given information pattern.

We will first study, in the next section, the soft-constrained linear-quadratic differential game, under OL, CLPS, SDPS and DPS information patterns, and present results on the boundedness of the upper value, existence of a saddle point, and characterization of saddle-point policies, all for the finite-horizon case. In this presentation, we will assume that the reader is familiar with the related theory on *conjugate points*, which has been included in Appendix A (Chapter 8). In Section 4.3, we will apply the results on the soft-constrained game to the disturbance attenuation problem (with $x_0 = 0$), under three information patterns, and will also include an illustrative example. These will then be extended to the infinite-horizon case in Section 4.4, and subsequently to the more general (original) problem (1.9)-(1.10) where the controlled output is given by (1.9b), and the initial state is not necessarily *zero*, but the information pattern is still CLPS (see Section 4.5). Section 4.6 discusses possible extensions of these results to nonlinear/nonquadratic problems of the type introduced by (2.13a) and (2.14), and finally the closing section (Section 4.7) summarizes the main point of the chapter.

4.2 The Soft-Constrained Differential Game

We study the soft-constrained linear-quadratic differential game described by (4.1) and (4.2b), under four different information patterns, where we take the initial state x_0 to be known, but not necessarily zero, as in Section 2.3.

4.2.1 Open-loop information structure for both players.

Here we can view the controls of the two players (Player 1 and Player 2) as elements of appropriate Hilbert spaces of square-integrable functions ($\mathcal{U} = \mathcal{H}_u$ and $\mathcal{W} = \mathcal{H}_w$, respectively), and the differential game becomes a static quadratic game defined on infinite dimensional spaces [19]. For the game to admit a unique saddle point, we again require the strict concavity of $L_\gamma(u,w)$ in $w \in \mathcal{W}$, for every fixed $u \in \mathcal{U}$. This is precisely the condition of existence of a unique solution to the LQ optimization problem

$$\max_{w \in \mathcal{W}} L_\gamma(u,w)$$

which is known to have an associated Riccati differential equation (see, Chapter 8, equation (8.4)). As discussed in Chapter 8, since the objective function is indefinite, this Riccati differential equation (RDE) may have "finite escape", in which case we say that "the Riccati equation has a *conjugate* point on the given interval". This now lays the ground for the following counterpart of Lemma 3.1.

Lemma 4.1. *For each fixed $\gamma > 0$, the quadratic objective functional $L_\gamma(u,w)$ given by (4.2b), and under the state equation (4.1), is strictly concave in $w \in \mathcal{W}$ for every open-loop policy $u \in \mathcal{U}$ of Player 1, if, and only if, the following Riccati differential equation does not have a conjugate point on the interval $[0, t_f]$:*

$$\dot{S} + A'S + SA + Q + \frac{1}{\gamma^2}SDD'S = 0; \quad S(t_f) = Q_f. \qquad (4.3)$$

Conversely, for each fixed finite interval $[0, t_f]$, there exists a $\hat{\gamma}^{\text{OL}} > 0$ such that the RDE (4.3) has no conjugate point on $[0, t_f]$ (hence L_γ is strictly concave in w) if $\gamma > \hat{\gamma}^{\text{OL}}$, and it has a conjugate point (and hence L_γ is not strictly concave in w) for all $\gamma \leq \hat{\gamma}^{\text{OL}}$.

Proof. See Propositions 8.1 - 8.3, in Chapter 8. ◇

Note that, for each fixed interval $[0, t_f]$, the threshold value, $\hat{\gamma}^{\mathrm{OL}}$, introduced in Lemma 4.1 above can be defined as

$$\hat{\gamma}^{\mathrm{OL}} := \inf\{\gamma > 0 : \text{(4.3) does not have a conjugate point on } [0, t_f]\}. \tag{4.4}$$

In fact, as shown in Proposition 8.3, the Riccati differential equation does have a conjugate point for $\gamma = \hat{\gamma}^{\mathrm{OL}}$.

Now, for each fixed $w \in \mathcal{W}$ and $x_0 \in \mathbb{R}^n$, the optimization problem

$$\min_{u \in \mathcal{U}} L_\gamma(u, w)$$

admits a unique solution, independently of γ, since it is quadratic and strictly convex in u. Under the conjugate-point condition of Lemma 4.1, on the other hand, the other optimization problem

$$\max_{w \in \mathcal{W}} L_\gamma(u, w)$$

admits a unique solution, since then we have strict concavity. Furthermore, both these solutions are linear in their arguments. Certain manipulations, details of which can be found in ([17], pp. 282 - 284) lead to the conclusion that there exists a unique fixed point of these two linear equations, which can be expressed in terms of a matrix function that is generated by another (than (4.3)) Riccati differential equation. The result is given below in Theorem 4.1, which can be viewed as the counterpart of Theorem 3.1 for the continuous-time case. In stating the necessary and sufficient conditions of this theorem, we find it more convenient to keep the interval $[0, t_f]$ fixed, and have the parameter γ as the variable, which will also be more relevant to the disturbance attenuation problem.

Theorem 4.1. *For the linear-quadratic differential game with open-loop information structure introduced above, let $\hat{\gamma}^{\mathrm{OL}}$ be as defined by (4.4), and* introduce the following Riccati differential equation:

$$\dot{Z} + A'Z + ZA + Q - Z(BB' - \frac{1}{\gamma^2}DD')Z = 0; \quad Z(t_f) = Q_f. \quad (4.5)$$

Then,

(i) For $\gamma > \hat{\gamma}^{\text{OL}}$, the RDE (4.5) does not have a conjugate point on the interval $[0, t_f]$.

(ii) For $\gamma > \hat{\gamma}^{\text{OL}}$, the game admits a unique saddle-point solution, given by

$$u^*(t) = \mu^*(t; x_0) = -B'(t)Z(t)x^*(t) \quad (4.6a)$$

$$w^*(t) = \nu^*(t; x_0) = \frac{1}{\gamma^2}D'(t)Z(t)x^*(t), \quad t \geq 0 \quad (4.6b)$$

where $x^*_{[0,t_f]}$ is the corresponding state trajectory, generated by

$$\dot{x}^* = (A - (BB' - \frac{1}{\gamma^2}DD')Z(t))x^*; \quad x^*(0) = x_0. \quad (4.7)$$

(iii) For $\gamma > \hat{\gamma}^{\text{OL}}$, the saddle-point value of the game is

$$L^*_\gamma = L_\gamma(u^*, w^*) = x'_0 Z(0) x_0. \quad (4.8)$$

(iv) If $\gamma < \hat{\gamma}^{\text{OL}}$, the upper value of the game is unbounded.

Proof. For parts (i) - (iii), see ([17], pp. 282-284); for part (iv), see Theorem 8.3 in Appendix A. ◇

Note that, implicit in the statement of part (i) of the Theorem is the fact that nonexistence of a conjugate point for (4.3) is a more stringent condition than the nonexistence of a conjugate point to (4.5). This fact can be demonstrated through a simple scalar example ([17], p. 292) as follows. Let the system dynamics be

$$\dot{x} = 2u - w$$

and the performance index be given by

$$L_1(u,w) = |x(t_f)|^2 + \int_0^{t_f} [|u(t)|^2 - |w(t)|^2]dt.$$

The Riccati differential equation (4.5) admits the unique solution

$$Z(t) = \frac{1}{1+t_f-t},$$

which is valid for all $t_f > 0$. Hence, there is no conjugate point for (4.5) on the interval $(0,\infty)$. For (4.3), however, a finite solution exists only if $t_f < 1$, in which case

$$S(t) = \frac{1}{1+t-t_f}, \quad 0 \le t \le t_f < 1.$$

Hence, (4.3) has a conjugate point if the interval is of unit length or longer.

4.2.2 Closed-loop perfect state information for both players.

Here, as in the discrete-time case, there exist generally multiple saddle-point equilibria, with again a refinement being possible by invoking *strong time consistency* or additive (Wiener) *noise-insensitivity*. As discussed in Section 2.3, both these refinement schemes lead to saddle-point policies that do not depend on memory (that is, on the past values of the state). In Section 2.3, we have called such a solution a *feedback* saddle-point solution. Any such solution is generated from the *Isaacs'* equation (2.16), which involves, for the linear-quadratic differential game, the solution of quadratic games pointwise in time. Now, if the Riccati differential equation (4.5) admits a continuously differentiable solution on the given interval (that is, if it does not have a conjugate point), then it is a simple exercise to show that the candidate *value function*

$$V(t;x) = x'Z(t)x, \quad t \in [0,t_f]$$

provides a solution to the partial differential equation (2.16). A corresponding pair of linear feedback saddle-point policies, which we will shortly present, can also readily be obtained from (2.16).

An even simpler, and more appealing, derivation of the feedback saddle-point solution, though, involves the "completion of squares" argument of quadratic optimization, applied to the LQ differential game. A direct computation shows that, under (4.1), the objective function L_γ given by (4.2a) can equivalently be written as

$$L_\gamma(u,w) = x_0' Z(0) x_0 + \int_0^{t_f} |u + B(t)'Z(t)x(t)|^2 dt \\ - \int_0^{t_f} |w - \frac{1}{\gamma^2} D(t)'Z(t)x(t)|^2 \gamma^2 dt, \quad (4.9)$$

provided (again) that, for the given $\gamma > 0$, there exists a continuously differentiable matrix function $Z(\cdot)$ solving (4.5) for all $t \in (0, t_f]$. Now, as in (4.4), let us introduce

$$\hat{\gamma}^{\mathrm{CL}} := \inf\{\gamma > 0 : \text{The RDE (4.5) does not have} \\ \text{a conjugate point on } [0, t_f]\}. \quad (4.10)$$

Hence, the equivalent representation (4.9) for the game performance index is valid whenever $\gamma > \hat{\gamma}^{\mathrm{CL}}$. Since this representation for L_γ completely decouples the controls of the players, it readily leads to the feedback saddle-point solution, by minimization and maximization of the second and third terms separately. This observation, together with the results presented in Appendix A, leads to the following theorem:

Theorem 4.2. *For the linear-quadratic zero-sum differential game with closed-loop information structure, defined on the time interval $[0, t_f]$, let the parameter $\hat{\gamma}^{\mathrm{CL}}$ be as defined by (4.10). Then:*

(i) If $\gamma > \hat{\gamma}^{\mathrm{CL}}$, the differential game admits a unique feedback saddle-point solution, which is given by

$$\mu^*(t; x(t)) = -B(t)'Z(t)x(t) \quad (4.11a)$$

$$\nu^*(t; x(t)) = \frac{1}{\gamma^2} D(t)'Z(t)x(t), \quad t \geq 0 \quad (4.11b)$$

where $Z(\cdot)$ is the unique solution of (4.5). Furthermore, the saddle-point value is given by (4.8), and the corresponding state trajectory is generated by (4.7).

(ii) If $\gamma < \hat{\gamma}^{\mathrm{CL}}$, the differential game has an unbounded upper value for all $\mu \in \mathcal{M}_{\mathrm{CLPS}}$.

Proof. Part (i) readily follows from the equivalent form (4.9) for the performance index. For a proof of part (ii), see Appendix A, and in particular Theorem 8.5. ◇

As in the discrete-time case, the implication of the last statement of the theorem is that the feedback saddle-point solution requires the least stringent conditions on the parameters of the game, and hence saddle-point solutions under other information structures (provided that they exist) can be constructed from the feedback saddle-point solution (4.11), on the state trajectory which is the solution of (4.7).[1] The open-loop solution of Theorem 4.1 is one such representation, where for existence (and boundedness of the upper value) we have required the additional condition given in Lemma 4.1. To obtain yet another representation of the feedback saddle-point solution, let Player 1's control policy depend not only on the current value of the state, but also on the past values of the control of Player 2. Further let $\mathcal{L}$ be a linear strictly causal mapping from $\mathcal{H}_w$ into $\mathcal{H}_u$, which further satisfies the *contraction* property:

$$\|\mathcal{L}(v)\|_u \leq \lambda \|v\|_w, \quad \forall v \in \mathcal{H}_w, \quad \text{for some } \lambda \in [0,1). \tag{4.12a}$$

Then, a representation of the feedback policy (4.11a) on the saddle-point trajectory is the memory strategy

$$u(t) = \mu(t; x(t), w_{[0,t)}) = -B'Zx(t) + \gamma[\mathcal{L}(w - \gamma^{-2}D'Zx)](t), \tag{4.12b}$$

---
[1] Again, recall the important feature of multiple saddle-point solutions, given in Theorem 2.5.

which (by Theorem 2.5) is a candidate saddle-point policy, along with (4.11b). Substitution of (4.12b) into (4.9), and maximization of the resulting expression with respect to $w \in \mathcal{H}_w$ (in view of the contraction property (4.12a) of $\mathcal{L}$) readily yield $x_0' Z(0) x_0$ as the maximum value, and hence the conclusion that (4.12b) is indeed a saddle-point policy whenever $\gamma > \hat{\gamma}^{\text{CL}}$. This solution, however, is neither strongly time consistent, nor noise insensitive.

Two other such constructions, which follow similar lines as above, are given below in Sections 4.2.3 and 4.2.4, in the contexts of sampled-data and delayed state information patterns.

4.2.3 Sampled-data information for both players.

We now endow both players with sampled-data perfect state (SDPS) information of the type introduced in Section 2.3. Note that the admissible controls for the players are of the form

$$u(t) = \mu(t; x(t_k), \ldots, x(t_1), x_0), \quad t_k \leq t < t_{k+1}, \quad k = 0, 1, \ldots, K, \quad (4.13a)$$

$$w(t) = \nu(t; x(t_k), \ldots, x(t_1), x_0), \quad t_k \leq t < t_{k+1}, \quad k = 0, 1, \ldots, K, \quad (4.13a)$$

where $\{0, t_1, t_2, \ldots, t_K\}$ is an increasing sequence of sampling times. In view of part (ii) of Theorem 4.2, together with Theorem 2.5, every sampled-data solution to the differential game has to be a *representation* of the feedback saddle-point solution (4.11) on the trajectory (4.7). One such representation, which is in fact the only one that is noise insensitive, and depends only on the most recent sampled value of the state, is

$$\mu^{SD}(t; x(t_k)) = -B(t)' Z(t) \Phi(t, t_k) x(t_k), \quad t_k \leq t < t_{k+1}, \quad (4.14a)$$

$$\nu^{SD}(t; x(t_k)) = \frac{1}{\gamma^2} D(t)' Z(t) \Phi(t, t_k) x(t_k), \quad t_k \leq t < t_{k+1}, \quad (4.14b)$$

where Φ is the state transition function associated with the matrix

$$F(t) := A(t) - [B(t)B(t)' - \frac{1}{\gamma^2} D(t) D(t)'] Z(t) \quad (4.14c)$$

and $Z(\cdot)$ is the solution to (4.5). Clearly, for (4.14a)-(4.14b) to constitute a saddle-point solution, first we have to require that (4.5) has no conjugate points in the given interval, and in addition we have to make sure that the function $J_\gamma(\mu^{SD}, w)$ is concave in $w \in \mathcal{H}_w$. The corresponding condition is given in Theorem 4.3 below, which also says that the solution (4.14a)-(4.14b) in fact requires the least stringent conditions, besides being strongly time consistent and noise insensitive.

Before presenting the theorem, let us introduce K Riccati differential equations, similar to (4.3) but with different boundary conditions:

$$\dot{S}_k + A'S_k + S_k A + Q + \gamma^{-2} S_k DD' S_k = 0;$$
$$S_k(t_{k+1}) = Z(t_{k+1}); \quad t_k \leq t < t_{k+1}, \quad k = K, K-1, \ldots 0. \quad (4.15)$$

Furthermore, let

$$\hat{\gamma}^{SD} := \inf\{\gamma > \hat{\gamma}^{CL} : \text{The } K \text{ RDE's (4.15) do not have conjugate}$$
$$\text{points on the respective sampling intervals}\}, \quad (4.16)$$

where $\hat{\gamma}^{CL}$ was defined by (4.10).

Then we have:

Theorem 4.3. *Consider the linear-quadratic zero-sum differential game (4.1)-(4.2a), with sampled-data state information (4.4).*

(i) *If $\gamma > \hat{\gamma}^{SD}$, it admits a unique strongly time-consistent and noise-insensitive saddle-point solution, which is given by the policies (4.14a)-(4.14b). The saddle-point value is again given by (4.8).*

(ii) *If $\gamma < \hat{\gamma}^{SD}$, the differential game with the given sampling times does not admit a saddle point, and it has unbounded upper value.*

Proof. We should first remark that the set of conjugate point conditions on the Riccati differential equations (4.15), which was used in defining $\hat{\gamma}^{SD}$, is precisely the condition for strict concavity of the function $J_\gamma(\mu^{\rm SD}, w)$ in $w \in \mathcal{H}_w$, which can be shown using Lemma 4.1. But simply showing this would not prove the theorem, since the policy (4.16a) was just one particular representation of the feedback saddle-point policy (4.11a). A more complete proof turns out to be one that follows basically the steps of the proof of Theorem 3.2, with the only difference being that in the present case the static games in between measurement points are continuous-time open-loop games (of the type covered by Theorem 4.1), while in the discrete-time game of Theorem 3.2 they were games defined on Euclidean spaces. Accordingly, under the sampled-data measurement scheme, we have the dynamic programming recursion (*Isaacs' equation*):

$$\min_{\mu \in \mathcal{M}_{\rm SDPS}} \max_{\nu \in \mathcal{N}_{\rm SDPS}} J_\gamma(\mu, \nu) = \min_{\mu_{[0,t_1)}} \max_{\nu_{[0,t_1)}} \left\{ \int_0^{t_1} (|x|^2_{Q(t)} + |u|^2 - \gamma^2 |w|^2) dt \right.$$

$$+ \ldots \min_{\mu_{[t_{K-1}, t_f]}} \max_{\nu_{[t_{K-1}, t_f]}} \left\{ \int_{t_{K-1}}^{t_f} (|x|^2_{Q(t)} + |u|^2 - \gamma^2 |w|^2) dt + |x(t_f)|^2_{Q_f} \right\} \right\}$$

where, on the time interval $[t_k, t_{k+1})$,

$$u(t) = \mu_{[t_k, t_{k+1})}(t, x(t_k), \ldots, x(t_1), x_0)$$

and

$$w(t) = \nu_{[t_k, t_{k+1})}(t, x(t_k), \ldots, x(t_1), x_0).$$

This dynamic programming recursion involves the solution of a sequence of (nested) open-loop linear-quadratic zero-sum differential games, each one having a bounded upper value if, and only if, (from Theorem 4.1) the Riccati differential equation (4.15) does not have a conjugate point on the corresponding subinterval. The boundary values are determined using Theorem 2.5, that is the fact that open-loop and closed-loop feedback saddle-point policies generate the same trajectory whenever both saddle points exist. It is also because of this reason that the saddle-point policies have the structure (4.15), which are also the unique strongly time-consistent and noise-insensitive saddle-point policies since they are derived using dynamic programming (see the discussion in Section 2.3). This completes the proof of part (i). Part (ii) follows from a reasoning identical to the one used in the proof of Theorem 3.2 (iii), under the observation (using Theorem 4.1) that if the upper value of any one of the open-loop games encountered in the derivation above is unbounded, then the upper value of the original game will also be unbounded with sampled-data measurements. ◇

Remark 4.1. In construction of the sample-data saddle-point strategies (4.15), we need the solution of a single n-dimensional RDE (4.5). However, to determine the existence of a saddle point we need, in addition, the solutions (or the conjugate point conditions) of K n-dimensional RDE's (4.15), one for each sampling subinterval. These conditions would still be the prevailing ones if Player 2 had access to full (continuous) state information, and the pair of policies (4.15) would still be in saddle-point equilibrium under this extended (closed-loop) information for Player 2. However, this solution will not retain its noise-insensitivity property. Under the CLPS information pattern for Player 2 (and still with SDPS pattern for Player 1), a noise-insensitive saddle-point policy for Player 2 will depend on the current value of the state, as well as the most recent sampled state measurements. For the construction of such a policy, which will again exist whenever $\gamma > \hat{\gamma}^{SD}$, we refer the reader to [13]. ◇

Remark 4.2. Since $\mathcal{M}_{SDPS} \subset \mathcal{M}_{CLPS}$, it should be intuitively obvious that if the upper value of the game is unbounded under continuous state measurements, it should also be unbounded under sampled state measurements, but not necessarily *vice versa* because sampling continuous state measurements generally leads to a loss of information for Player 1. This intuition is of course also reflected in the definition of the parameter value $\hat{\gamma}^{SD}$, in (4.16). An important consequence of this reasoning is that, because of the loss of information due to sampling, for "reasonable" problems we might expect to see the strict inequality[2]:

$$\hat{\gamma}^{\mathrm{CL}} < \hat{\gamma}^{\mathrm{SD}}. \tag{4.17}$$

Derivation of the precise conditions under which this would be true for the time-varying problem is quite a challenging task, which we will not pursue here. ◇

4.2.4 Delayed state measurements

Here we consider the linear-quadratic differential game where both players have access to state with a delay of $\theta > 0$ time units. The players' admissible control policies will hence be of the form:

$$\begin{aligned} u(t) &= \mu(t, x_{[0,t-\theta]}), \quad t \geq \theta \\ &= \mu(t, x_0), \qquad 0 \leq t < \theta. \end{aligned} \tag{4.18}$$

for Player 1, and a similar structure for Player 2. We denote the class of such controllers for Players 1 and 2 by $\mathcal{M}_{\theta D}$ and $\mathcal{N}_{\theta D}$, respectively. Since $\mathcal{M}_{\theta D} \subset \mathcal{M}_{\mathrm{CLPS}}$, we again know from Theorem 4.2 that for $\gamma < \hat{\gamma}^{\mathrm{CL}}$ the upper value of this differential game (with delayed measurements) will be unbounded, and for $\gamma > \hat{\gamma}^{\mathrm{CL}}$ its saddle-point solution is necessarily a representation of the feedback solution (4.15), provided that it exists. Motivated by the argument that led to Theorem 3.4 in Chapter 3 for the discrete-time delayed-measurement game, we can now write down the following representation of (4.15) in $\mathcal{M}_{\theta D} \times \mathcal{N}_{\theta D}$ as a candidate saddle-point solution, which is clearly noise insensitive:

$$u(t) = \mu^\circ(t, \eta(t)) = -B(t)'Z(t)\eta(t), \quad 0 \leq t \leq t_f, \tag{4.19a}$$

$$w(t) = \nu^\circ(t, \eta(t)) = \frac{1}{\gamma^2} D(t)'Z(t)\eta(t), \quad 0 \leq t \leq t_f, \tag{4.19b}$$

---

[2]Note that if the game is completely decoupled in terms of the effects of the controls of Player 1 and Player 2 on the state vector as well as the cost function (such as matrices A and Q being block-diagonal and the range spaces of BB' and DD' having empty intersection), then we would have an equality in (4.17); this, however, is not a problem of real interest to us, especially in the context of the disturbance attenuation problem to be studied in the next section.

where $\eta(\cdot)$ is generated by the (infinite-dimensional) compensator

$$\eta(t) = \psi(t, t-\theta)x(t-\theta) - \int_{t-\theta}^{t} \psi(t,s)$$

$$\cdot \left[B(s)B(s)' - \frac{1}{\gamma^2} D(s)D(s)' \right] Z(s)\eta(s) ds ,$$
(4.20)

$$\eta(s) = 0, \quad 0 \le s < \tau,$$

with $\psi(\cdot, \cdot)$ being the state transition matrix associated with $A(\cdot)$. To obtain the conditions under which this is indeed a saddle-point solution, we have to substitute (4.19a) into L_γ, and require the resulting quadratic function (in w) to be strictly concave. The underlying problem is a quadratic maximization problem in the Hilbert space $\mathcal{H}_w$, and is fairly complex because the state dynamics are described by a delay-differential equation. However, an alternative argument that essentially uses the property given in Theorem 2.5, yields the required concavity condition quite readily:

> If the problem admits a saddle-point solution, at any time τ the value of the game for the remaining future portion will be $x(\tau)'Z(\tau)x(\tau)$, which follows from Theorem 4.2 and Theorem 2.5. In the interval $[\tau-\theta, \tau]$, the choice of $\mu_{[\tau-\theta,\tau]}$ will not show any dependence on $w_{[\tau-\theta,\tau]}$, and hence the game will require essentially the same existence conditions as the open-loop game of Theorem 4.1, only defined on a shorter interval.

Hence, for each $\tau \ge \theta$, we will have a RDE of the type (4.3), defined on $[\tau-\theta, \tau]$:

$$\dot{S}_\tau + A'S_\tau + S_\tau A + Q + \gamma^{-2} S_\tau DD' S_\tau = 0;$$

$$S_\tau(\tau) = Z(\tau); \quad \tau-\theta \le t < \tau, \quad \tau \in [\theta, t_f].$$
(4.21)

This is a continuum family of Riccati differential equations, each one indexed by the time point $\tau \ge \theta$. Associated with this family, we now introduce the counterpart of (4.16):

$$\hat{\gamma}^{\theta D} := \inf\{\gamma > \hat{\gamma}^{\mathrm{CL}} : \text{For every } \tau \geq \theta, \text{ the RDE (4.21) does not have a conjugate point on the corresponding interval}\}. \tag{4.22}$$

It now follows from the discussion above that the delayed-measurement game cannot have a bounded upper value if $\gamma < \hat{\gamma}^{\theta D}$. Hence, we arrive at the following counterpart of Theorem 4.3 under delayed measurements.

Theorem 4.4. *Consider the linear-quadratic zero-sum differential game (4.1)-(4.2a), with delayed state measurements (4.18). Let the scalar $\hat{\gamma}^{\theta D}$ be defined by (4.22). Then:*

(i) *If $\gamma > \hat{\gamma}^{\theta D}$, the game admits a unique noise-insensitive saddle-point solution, which is given by the policies (4.19a)-(4.19b). The saddle-point value is again given by (4.8).*

(ii) *If $\gamma < \hat{\gamma}^{\theta D}$, the differential game does not admit a saddle point and has unbounded upper value.*

◇

4.3 The Disturbance Attenuation Problem

We now return to the minimax design problem formulated in Chapter 1, by (1.9)-(1.11), where we take $H'G = 0$, $H'H = Q$, $G'G = I$, $C = I$, $E = 0$ and $x_0 = 0$, or equivalently the problem formulated in Section 4.1 by (4.1)-(4.2a), with the H^∞ norm to be minimized given by (1.2b). This is the disturbance attenuation problem with closed-loop perfect state information (2.15a), where the controller space is denoted by $\mathcal{M}_{\mathrm{CL}}$. We will also consider the sampled-data state information (2.15b), with the resulting controller belonging to $\mathcal{M}_{\mathrm{SD}}$, and the delayed state information (2.15c), where the controller belongs to $\mathcal{M}_{\theta D}$. In all cases, the solutions to the disturbance attenuation problem (to be presented below) follow from the solutions of the corresponding soft-constrained games studied in the previous section – – a relationship which has already been expounded on in general terms in Section 1.2, and in the context of the discrete-time problem in Section 3.3.

4.3.1 Closed-loop perfect state information

The associated soft-constrained game is the one studied in Section 4.2.2. Let $\hat{\gamma}^{\mathrm{CL}}$ be as defined by (4.10). For every $\gamma > \hat{\gamma}^{\mathrm{CL}}$, we know (from Theorem 4.2) that the soft-constrained game admits a saddle-point solution, with the minimizing control given by (4.11a). Let us rewrite this feedback controller as μ_γ^*, to indicate explicitly its dependence on the parameter γ. Unlike the case of the discrete-time problem (cf. Theorem 3.5) the limit of μ_γ^* as $\gamma \downarrow \hat{\gamma}^{\mathrm{CL}}$ may not be well-defined (because of the existence of a conjugate point to (4.5) on the interval $[0, t_f]$), and hence we have to be content with suboptimal solutions. Toward a characterization of one such solution, let $\epsilon > 0$ be sufficiently small, and $\gamma_\epsilon := \hat{\gamma}^{\mathrm{CL}} + \epsilon$. Then, it follows from the saddle-point value (4.8), by taking $x_0 = 0$, that

$$\|T_{\mu_{\gamma_\epsilon}^*}(w)\|^2 \equiv J(\mu_{\gamma_\epsilon}^*, w) \le \gamma_\epsilon^2 \|w\|_w^2, \quad \forall w \in \mathcal{H}_w.$$

This implies, in view of part (ii) of Theorem 4.2, that

$$\hat{\gamma}^{\mathrm{CL}} = \gamma^* \ (\equiv \gamma^{\mathrm{CL}}),$$

where the latter is the optimum (minimax) disturbance attenuation level (1.2a) under closed-loop state information. Hence we arrive at the following theorem:

Theorem 4.5. *For the continuous-time finite-horizon disturbance attenuation problem with closed-loop perfect state information (and with $x_0 = 0$):*

(i) The minimax attenuation level is equal to $\hat{\gamma}^{\mathrm{CL}}$, that is

$$\gamma^* := \inf_{\mu \in \mathcal{M}_{\mathrm{CLPS}}} \ll T_\mu \gg = \hat{\gamma}^{\mathrm{CL}} \qquad (4.23a)$$

where the bound is given by (4.10).

(ii) Given any $\epsilon > 0$,
$$\ll T_{\mu^*_{\gamma_\epsilon}} \gg \ \leq \gamma_\epsilon := \hat{\gamma}^{\mathrm{CL}} + \epsilon, \qquad (4.23b)$$
where the suboptimal controller $\mu^*_{\gamma_\epsilon}$ that achieves this bound is given by
$$\mu^*_{\gamma_\epsilon}(t; x(t)) = -B(t)' Z_{\gamma_\epsilon}(t) x(t), \qquad (4.23c)$$
where $Z_{\gamma_\epsilon}(\cdot)$ is the unique nonnegative definite solution of the RDE (4.5), with $\gamma = \gamma_\epsilon$.

(iii) If the controller is allowed to depend also on the past values of the disturbance, a class of controllers that achieve the same bound (4.23b) is given by
$$u(t) = \mu_{\gamma_\epsilon}(t; x(t), w_{[0,t)}) = -B' Z_{\gamma_\epsilon} x(t) + \gamma_\epsilon [\mathcal{L}(w - \gamma_\epsilon^{-2} D' Z_{\gamma_\epsilon} x)](t), \qquad (4.24)$$
where $\mathcal{L}$ is a linear contraction operator, as introduced by (4.12a).

Proof. Parts (i) and (ii) follow readily from Theorem 4.2, in view of the discussion preceding the statement of the theorem. Note that, it is not possible for $\gamma^* < \hat{\gamma}^{\mathrm{CL}}$, because for $\gamma < \hat{\gamma}^{\mathrm{CL}}$ the upper value of the related soft-constrained game is unbounded by Theorem 4.2 (ii). Finally, part (iii) of the theorem follows from (4.12b), and the discussion following Theorem 4.2. ◇

Remark 4.3. Note that, contrary to the discrete-time case covered by Theorem 3.5, Theorem 4.5 does not claim existence of an optimal controller to the continuous-time disturbance attenuation problem. The reason for this is that the sequence $\{\mu^*_{\gamma_\epsilon}, \epsilon > 0\}$ may not admit a well-defined limit in $\mathcal{M}_{\mathrm{CLPS}}$ as $\epsilon \downarrow 0$, because of the existence of a conjugate point (finite escape) for the RDE (4.5). This will be demonstrated on an example in Section 4.3.3. ◇

4.3.2 Sampled state measurements

We present here the counterpart of Theorem 4.5 for the sampled-data measurement case, where the relevant soft-constrained game is the one covered by Theorem 4.3. The main result in this context is the equality

$$\gamma^{\text{SD}} = \hat{\gamma}^{\text{SD}}, \tag{4.25}$$

where γ^{SD} is the minimax attenuation level under sampled-data measurement (the counterpart of (1.2a), and $\hat{\gamma}^{\text{SD}}$ is given by (4.16). To prove this result, we follow a line of reasoning similar to that used in Section 4.3.1. Let $\epsilon > 0$ be sufficiently small, and $\hat{\gamma}_\epsilon := \hat{\gamma}^{\text{SD}} + \epsilon$. Then, it follows from Theorem 4.3, and using the saddle-point controller (4.14a) with $\gamma = \hat{\gamma}_\epsilon$ and the saddle-point value (4.8), that

$$J(\mu_{\hat{\gamma}_\epsilon}^{\text{SD}}, w) \leq x_0' Z_{\hat{\gamma}_\epsilon}(0) x_0, \quad \text{for all } w \in \mathcal{H}_w,$$

where both $\epsilon > 0$ and $x_0 \in \mathbb{R}^n$ are arbitrary. This implies, by taking $x_0 = 0$, that

$$\|T_{\mu_{\hat{\gamma}_\epsilon}^{\text{SD}}}(w)\| \leq \hat{\gamma}_\epsilon \|w\|, \quad \text{for all } w \in \mathcal{H}_w,$$

leading to the inequality

$$\gamma^{\text{SD}} \leq \hat{\gamma}^{\text{SD}} + \epsilon.$$

To complete the proof of the desired equality, we now show that the strict inequality $\gamma^{\text{SD}} < \hat{\gamma}^{\text{SD}}$ is not possible. Suppose this strict inequality were true. Then we would have, for some $\mu \in \mathcal{M}_{\text{SD}}$, say $\tilde{\mu}$,

$$\ll T_{\tilde{\mu}}^\gamma \gg < \hat{\gamma}^{\text{SD}} \implies \sup_{w \in \mathcal{H}_w} J_{\gamma^{\text{SD}}}(\tilde{\mu}, w) < 0,$$

which says that the upper value of the soft-constrained game (under sampled state measurements) is bounded, in spite of the fact that $\gamma < \hat{\gamma}^{\text{SD}}$. This contradicts with statement (ii) of Theorem 4.3, thus completing the proof.

We are now in a position to state the following theorem, which follows readily from Theorem 4.3, in view of (4.25):

Theorem 4.6. *For the finite-horizon disturbance attenuation problem with sampled state information (and $x_0 = 0$):*

(i) The minimax attenuation level γ^{SD} is equal to $\hat{\gamma}^{SD}$ given in (4.16).

(ii) Given any $\epsilon > 0$, and a corresponding $\hat{\gamma}_\epsilon := \hat{\gamma}^{SD} + \epsilon$, the controller given by (4.14a), that is

$$\mu^{SD}_{\hat{\gamma}_\epsilon}(t; x(t_k)) = -B(t)' Z_{\hat{\gamma}_\epsilon}(t) \Phi_{\hat{\gamma}_\epsilon}(t, t_k) x(t_k), \quad t_k \leq t < t_{k+1}, \tag{4.26a}$$

achieves a performance no worse than $\hat{\gamma}_\epsilon$. In other words,

$$\|T_{\mu^{SD}_{\hat{\gamma}_\epsilon}}(w)\| \leq \hat{\gamma}_\epsilon \|w\|, \quad \text{for all } w \in \mathcal{H}_w. \tag{4.26b}$$

(iii) if $\gamma^{SD} > \gamma^*$, the limiting controller

$$\mu^{SD}_{\hat{\gamma}^{SD}} = \lim_{\epsilon \downarrow 0} \mu^{SD}_{\hat{\gamma}_\epsilon} \tag{4.26c}$$

exists.

Proof. Parts (i) and (ii) have already been proven prior to the statement of the theorem. To prove part (iii), it is sufficient to observe that under the given condition the RDE (4.5) does not have a conjugate point on $[0, t_f]$, for $\gamma = \hat{\gamma}^{SD}$, and hence the controller $\mu^{SD}_{\hat{\gamma}_\epsilon}$ is well defined at $\epsilon = 0$. ◇

4.3.3 An illustrative example

The following scalar example will serve to illustrate the general set of results presented in Theorems 4.5 and 4.6.

Consider the scalar plant

$$\dot{x} = u + w, \quad x(0) = 0, \quad t \geq 0,$$

along with the performance index

$$\int_0^{t_f} ([x(t)]^2 + [u(t)]^2) dt.$$

Under perfect state measurements, the minimax attenuation level is determined by the conjugate-point condition of the RDE

$$\dot{Z} + 1 - \left(1 - \frac{1}{\gamma^2}\right) Z^2 = 0; \quad Z(t_f) = 0 \tag{*}$$

whose unique solution, for $\gamma^2 < 1$, is

$$Z_\gamma(t) = \frac{1}{m} \tan[m(t_f - t)], \quad 0 \le t \le t_f$$

$$m := \sqrt{(1-\gamma^2)/\gamma^2}$$

provided that $0 < mt_f < \frac{\pi}{2}$. Hence, the minimax disturbance attenuation level is

$$\gamma^* = 2t_f / \sqrt{\pi^2 + 4t_f^2}.$$

Given any $\epsilon > 0$, and the corresponding $\gamma_\epsilon := \gamma^* + \epsilon$, the controller

$$\mu^*_{\gamma_\epsilon}(t, x(t)) = -Z_{\gamma_\epsilon}(t)x(t), \quad 0 \le t \le t_f$$

achieves an attenuation level that is no higher than γ_ϵ. Note that this becomes a "high-gain" controller as $\epsilon \downarrow 0$.

To obtain some explicit (numerical) results for the sampled-state measurement scheme, let us take $t_f = \pi$, which leads to $\gamma^* = 2/\sqrt{5}$. Now let there be a single sampling point, at $t = (1-\lambda)\pi$, where $0 < \lambda < 1$. Anticipating an attenuation level $\gamma^{SD} > 1$, we first solve $(*)$ to obtain

$$Z_\gamma(t) = \frac{1}{\sigma} \tanh[\sigma(t_f - t)] \equiv \frac{1}{\sigma} \tanh[\sigma(\pi - t)]$$

$$\sigma := \sqrt{(\gamma^2 - 1)/\gamma^2}$$

which provides the boundary condition to (3.33) at the sampling time $t = (1-\lambda)\pi$. The conjugate-point conditions in the intervals $[(1-\lambda)\pi, \pi]$ and $[0, (1-\lambda)\pi]$ dictate, respectively, the following inequalities:

i) $\lambda\pi < \frac{\pi}{2}\gamma$.

ii) $(1-\lambda)\pi + \gamma \arctan\left[\frac{1}{\sqrt{\gamma^2-1}}(\tanh[\lambda\pi\sqrt{\gamma^2-1}/\gamma])\right] < \frac{\pi}{2}\gamma$

for $\gamma > 1$, which indeed turns out to be the case because it can be shown that for $\gamma < 1$, regardless of the value of $\lambda \in (0,1)$, either i) or ii) is violated. At this point we can raise the question of the "optimal choice" of the sampling time parameter λ, so that γ^{SD} is minimized. Because of the monotonicity property of the two conditions above, the optimum value of λ is one under which there exists a γ which makes both i) and ii) equalities (simultaneously). Some manipulations bring this condition down to one of existence of a $\lambda \in (0,1)$ to the trigonometric equation

$$\tan\left[\frac{(2\lambda - 1)}{\lambda} \cdot \frac{\pi}{2}\right] = \frac{1}{4\lambda^2 - 1} \tanh\left[\frac{\pi}{2}\sqrt{4\lambda^2 - 1}\right]$$

which admits the unique solution

$$\lambda \approx 0.6765.$$

The corresponding minimax attenuation level is

$$\gamma^{\text{SD}} = 2\lambda \approx 1.353.$$

Note that the above also shows that in the general disturbance attenuation problem if the choice of sampling times is also part of the design, under the constraint of a fixed total number of sampled measurements in the given time interval, uniform sampling (*i.e.*, $t_{k+1} - t_k = $ constant) will not necessarily be optimal. We will later see that for the time-invariant infinite-horizon problem uniform sampling is indeed optimal.

4.3.4 Delayed state measurements

Consider now the case when the controller can access the state with a delay of $\theta > 0$ time units. The soft-constrained game here is the one discussed in Section 4.2.4, whose relevant saddle-point solution was given in Theorem 4.4. An argument quite identical to the one used in the previous section, applied now to Theorem 4.4, readily leads to the following result.

Theorem 4.7. *Consider the continuous-time disturbance attenuation problem with θ-delayed state information, of the type given by (4.18). Let the scalar $\hat{\gamma}^{\theta D}$ be defined by (4.22). Then:*

(i) The minimax attenuation level is equal to $\hat{\gamma}^{\theta D}$, that is

$$\gamma^\circ := \inf_{\mu \in \mathcal{M}_{\theta D}} \ll T_\mu \gg \; = \hat{\gamma}^{\theta D} \qquad (4.27a)$$

(ii) For any $\epsilon > 0$,

$$\ll T_{\mu^\circ_{\gamma^\circ_\epsilon}} \gg \; \leq \gamma^\circ_\epsilon := \gamma^\circ + \epsilon \qquad (4.27b)$$

where $\mu^\circ_{\gamma^\circ_\epsilon}$ is given by (4.19a), with $\gamma = \gamma^\circ_\epsilon$.

(iii) If $\gamma^\circ > \gamma^*$, where the latter was given in Theorem 4.5, the limiting controller

$$\mu^\circ_{\gamma^\circ} = \lim_{\epsilon \downarrow 0} \mu^\circ_{\gamma^\circ_\epsilon} \qquad (4.27c)$$

exists. ◇

4.4 The Infinite-Horizon Case

In this section we study the time-invariant versions of the problems covered in Sections 4.2 and 4.3, as $t_f \to \infty$, first for the soft-constrained game and then for the disturbance attenuation problem. For future reference, let us introduce the notation J^∞ and J^∞_γ to denote the infinite-horizon versions of the objective functions (4.2a) and (4.2b), respectively.

4.4.1 The soft-constrained differential game

Drawing a parallel with the analysis of Section 3.4 in the discrete-time case, we would expect here the *steady-state* feedback saddle-point solution for the soft-constrained perfect state information game to be in the form (from (4.11))

$$u^\infty(t) = \mu^\infty(x(t)) = -B'\bar{Z}x(t) \qquad (4.28a)$$

$$w^\infty(t) = \nu^\infty(x(t)) = \frac{1}{\gamma^2}D'\bar{Z}x(t), \quad t \geq 0, \qquad (4.28b)$$

where all matrices are constants, and $\bar{Z}$ is the limiting solution of (4.5), which (if exists) should satisfy the continuous-time algebraic Riccati equation (ARE)

$$A'\bar{Z} + \bar{Z}A + Q - \bar{Z}(BB' - \frac{1}{\gamma^2}DD')\bar{Z} = 0. \tag{4.29}$$

Denote the solution of the Riccati differential equation (4.5) by $Z(t;t_f)$, where dependence on the terminal time is explicitly recognized. Now, the first question that would be of interest is:

> If $Z(t;t_f)$ is bounded for all $t_f > 0$ (i.e., (4.5) does not have any conjugate points on $[0,t_f]$ for any t_f) and $\lim_{t_f \to \infty} Z(t,t_f) = \bar{Z}$ exists, does this necessarily imply that the pair (4.28) is in saddle-point equilibrium?

The answer is "no", as the following (counter) example, taken from [64] clearly illustrates.

Example 4.1. Consider the scalar state dynamics

$$\dot{x} = x + u + w, \quad x(0) = x_0$$

and the objective function

$$J_\gamma^\infty = \int_0^\infty (x^2 + u^2 - \gamma^2 w^2)dt, \quad \gamma^2 = 2.$$

The Riccati differential equation (4.5) associated with this soft-constrained differential game is

$$\dot{Z} + 2Z + 1 - \frac{1}{2}Z^2 = 0; \quad Z(t_f;t_f) = 0$$

which admits the unique solution

$$Z(t;t_f) = \sqrt{6}\tanh\left[\sqrt{6}(t_f - t) + \tanh^{-1}(-2\sqrt{6})\right] + 2.$$

This function has a well-defined limit, as $t_f \to \infty$, which is independent of t:

$$\bar{Z} = 2 + \sqrt{6}.$$

Hence the policies (4.28a)-(4.28b) are:

$$\mu^\infty(x) = -(2+\sqrt{6})x; \quad \nu^\infty(x) = \frac{2+\sqrt{6}}{2}x.$$

Note that the closed-loop system under these policies is

$$\dot{x} = -\frac{\sqrt{6}}{2}x$$

which is asymptotically stable. Also, the system under only μ^∞ is

$$\dot{x} = -(1+\sqrt{6})x + w,$$

which is bounded input–bounded output stable.

Now it is easy to see that if $u = \mu^\infty(x)$, the J^∞-maximizing solution for Player 2 is ν^∞. However, when $w = \nu^\infty(x)$, the J^∞-minimizing solution for Player 1 is not μ^∞. The underlying minimization problem then is

$$\int_0^\infty \left[-\left(4+2\sqrt{6}\right)x^2 + u^2\right] dt \to \text{minimum}$$

subject to

$$\dot{x} = \frac{4+\sqrt{6}}{2}x + u.$$

Since the open-loop system is unstable, and there is a negative cost on state in the objective function, clearly the minimizing solution is $u \equiv 0$, under which the cost is driven to $-\infty$.

Even though there is no continuity in the minimizer's policy at $t_f = \infty$, nevertheless the value (defined by (2.11b)) is continuous. It can be shown (see [64]) that given an $\epsilon > 0$, one can find a $T(\epsilon) > 0$ such that by choosing the time-varying feedback policy

$$\nu_\epsilon(t, x(t)) = \begin{cases} \gamma^{-2} D'Z(t;T)x(t), & 0 \le t < T \\ 0, & t \ge T \end{cases},$$

Player 2 can ensure that

$$\min_u J^\infty(u, \nu_\epsilon) \ge (2+\sqrt{6})(x_0)^2 - \epsilon(x_0)^2$$

where

$$(2+\sqrt{6})(x_0)^2 = \lim_{t_f \to \infty} Z(t;t_f)(x_0)^2$$

is the limit of the value of the finite horizon game (as $t_f \to \infty$). Since $\epsilon > 0$ can be chosen to be arbitrarily small, this shows that the infinite-horizon game of this example indeed has a value. ◇

Note that, in the above example, there is no continuity in the saddle-point property of the maximizer's feedback policy, as $t_f \to \infty$, whereas there is continuity in the value of the game and in the feedback policy of the minimizer. This feature actually turns out to be valid for the general linear-quadratic differential game,[3] under some general conditions, as stated in the following theorem.

a) **Closed-loop perfect state measurements :**

Theorem 4.8. *Consider the infinite-horizon time-invariant linear quadratic soft-constrained differential game where, in the formulation of Section 4.1, the information pattern is CLPS, $Q_f = 0$, and the pair (A, H) is observable, with $H'H = Q$. Then:*

(i) *For each fixed t, the solution to (4.5), $Z(t;t_f)$, is nondecreasing in t_f; that is, if (4.5) has no conjugate point in a given interval $[0,t_f]$,*

$$Z(t;t'_f) - Z(t;t''_f) \geq 0, \quad t_f > t'_f > t''_f \geq 0. \qquad (4.30)$$

(ii) *Every nonnegative definite solution of (4.29) is in fact positive definite, and if there exists a positive definite solution there is a minimal such solution, denoted $\bar{Z}^+$. This matrix has the property that $\bar{Z}^+ - Z(t;t_f) \geq 0$ for all $t_f \geq 0$.*

---

[3] This is, in fact, all we need for the disturbance attenuation problem, since our interest there is primarily in the boundedness of the upper value, and the construction of only minimizer's (Player 1's) saddle-point policy, as the limit of the feedback policies obtained from finite-horizon games.

(iii) The differential game has equal upper and lower values if, and only if, the ARE (4.29) admits a positive definite solution, in which case the common value is

$$J_\gamma^{\infty *} = x_0' \bar{Z}^+ x_0. \tag{4.31}$$

(iv) The upper value is finite if, and only if, the upper and lower values are equal.

(v) If $\bar{Z}^+ > 0$ exists, as given above, the steady-state feedback controller μ^∞ given by (4.28a), with $\bar{Z}$ replaced by $\bar{Z}^+$, attains the finite upper value, in the sense that

$$\sup_{\nu \in \mathcal{N}_{\text{CLPS}}} J_\gamma^\infty(\mu^\infty, \nu) = x_0' \bar{Z}^+ x_0, \tag{4.32}$$

and the maximizing feedback solution in (4.32) is given by (4.28b), again with $\bar{Z} = \bar{Z}^+$.

(vi) Whenever the upper value is bounded, the two feedback matrices

$$A - (BB' - \frac{1}{\gamma^2} DD') \bar{Z}^+ \tag{4.33a}$$

and

$$A - BB' \bar{Z}^+ \tag{4.33b}$$

are asymptotically stable.

Proof. The proof is similar to that of Theorem 3.7, and follows by constructing natural counterparts of Lemmas 3.3–3.6 in the continuous time. In order to avoid duplication, we do not provide the full details for all the cases here, but only verify item (v). We should note, at this point, that this theorem (with the exception of (4.33b)) was first presented in [64] for the case when Q is of full rank, where it was also stated that the result is equally valid if the condition $Q > 0$ is replaced by the observability of the pair (A, H).

Now, toward establishing (v), consider the finite-horizon truncation of the maximization problem (4.32):

$$\max_{w_{[0,t_f]}} \int_0^{t_f} (|x|_Q^2 + |B'\bar{Z}^+x|^2 - \gamma^2|w|^2)dt$$

subject to

$$\dot{x} = (A - BB'\bar{Z}^+)x + Dw \equiv \widetilde{A}x + Dw.$$

Since $\bar{Z}^+ > 0$, by hypothesis, this quantity is bounded from above by

$$\max_{w_{[0,t_f]}} \left\{ \int_0^{t_f} (|x|_{\widetilde{Q}}^2 - \gamma^2|w|^2)dt + |x(t_f)|_{\bar{Z}^+}^2 \right\},$$

where

$$\widetilde{Q} := Q + \bar{Z}^+ BB'\bar{Z}^+ .$$

In view of Lemma 4.1 and Theorem 8.2 of Appendix A, this maximization problem admits a unique solution provided that the following RDE does not have a conjugate point on $[0, t_f]$:

$$\dot{\widetilde{S}} + \widetilde{A}'\widetilde{S} + \widetilde{S}\widetilde{A} + \widetilde{Q} + \frac{1}{\gamma^2}\widetilde{S}DD'\widetilde{S} = 0; \quad \widetilde{S}(t_f) = \bar{Z}^+,$$

which, however, admits the solution

$$\widetilde{S} = \bar{Z}^+.$$

Furthermore, the maximum value is given by

$$\bar{x}_0' \bar{Z}^+ x_0,$$

and the maximizing solution is (4.28b) with $\bar{Z} = \bar{Z}^+$. Since the truncated maximization problem is uniformly bounded from above, with the bound being the value (4.31), the result stands proven. ◇

b) **Sampled-data perfect state information :**

Now, given that $\lim_{t_f \to \infty} Z(t; t_f) = \bar{Z}^+$, where $\bar{Z}^+$ exists as the minimal positive definite solution of the ARE (4.29), it is fairly straightforward to obtain the counterpart of Theorem 4.8 under sampled-data measurements. In view of Theorem 4.3, we have, as candidate solutions,

$$\mu_\infty^{SD}(t, x(t_k)) = -B'\bar{Z}^+\Phi(t, t_k)x(t_k), \quad t_k \leq t < t_{k+1} \tag{4.34a}$$

$$\nu_\infty^{SD}(t, x(t_k)) = \gamma^{-2}D'\bar{Z}^+\Phi(t, t_k)x(t_k), \quad t_k \leq t < t_{k+1} \tag{4.34b}$$

where Φ is the state transition function associated with the matrix

$$\bar{F} := A - \left(BB' - \frac{1}{\gamma^2}DD'\right)\bar{Z}^+. \tag{4.34c}$$

In between two sampling points, still the Riccati differential equations (4.15) will have to be used, for existence considerations, with the boundary conditions now being identical: $S(t_k) = \bar{Z}^+$. Being time invariant, the Riccati differential equations themselves are also identical, apart from the length of the time interval on which they are defined. Since these Riccati equations are used only for the determination of existence, the conjugate-point condition of the one defined on the longest time interval will be the determining factor -- all others will play no role in the solution. Hence, letting

$$\bar{t}_s := \sup_k \{t_{k+1} - t_k, \quad k = 0, 1, \ldots\}, \tag{4.35}$$

the relevant RDE will be

$$\dot{S} + A'S + SA + Q + \frac{1}{\gamma^2}SDD'S = 0; \quad S(\bar{t}_s) = \bar{Z}^+ \tag{4.36}$$

which should not have a conjugate point in the interval $[0, \bar{t}_s]$. It is now convenient to introduce here the counterpart of (4.16):

$$\hat{\gamma}_\infty^{SD} := \inf\{\gamma > 0 : \text{The ARE (4.29) admits a positive definite solution,}$$

and the RDE (4.36) does not have a conjugate point

on the interval $[0, \bar{t}_s]\}$.

$$\tag{4.37}$$

This brings us to the following theorem.

Theorem 4.9. *Consider the infinite-horizon time-invariant linear-quadratic differential game, where $Q_f = 0$, the pair (A, H) is observable, with $H'H = Q$, and the information structure is sampled-data. Then,*

(i) *If $\gamma > \hat{\gamma}_\infty^{\text{SD}}$, the game has equal upper and lower values, with the common value being (4.31a).*

(ii) *If $\gamma > \hat{\gamma}_\infty^{\text{SD}}$, the controller μ_∞^{SD} given by (4.34a) attains the finite value, that is*

$$\sup_{w \in \mathcal{H}_w} J_\gamma^\infty(\mu_\infty^{\text{SD}}, w) = x_0' \bar{Z}^+ x_0.$$

(iii) *If $\gamma < \hat{\gamma}_\infty^{\text{SD}}$, the upper value of the differential game is unbounded.*

(iv) *If $\gamma > \hat{\gamma}_\infty^{\text{SD}}$, the matrix Q is positive definite, and the controller (4.34a) is used for u, the resulting (hybrid) closed-loop system becomes bounded input–bounded output stable (with w considered to be the input).*

Proof. Parts (i) and (iii) follow from Theorem 4.8 and the discussion preceding the statement of this theorem (see also Theorem 4.3).

To prove part (iii), we use the line of reasoning employed in the proof of Theorem 4.8 (v). Note that the optimization problem in (ii) above is the maximization of the performance index

$$\limsup_{K \uparrow \infty} \sum_{k=0}^{K-1} \underbrace{\int_{t_k}^{t_{k+1}} (|x(t)|_Q^2 + |B'\bar{Z}^+ \Phi(t, t_k)|^2 - \gamma^2 |w(t)|^2) dt}_{F_k(x, w)}$$

over $w_{[0,\infty)}$, subject to the state equation constraint

$$\dot{x} = Ax + Dw - BB'\bar{Z}^+\Phi(t, t_k)x(t_k), \quad t_k \leq t < t_{k+1}.$$

First consider the truncated version with K sampling intervals:

$$\max_{w_{[0, t_k)}} \sum_{k=0}^{K-1} F_k(x, w) \qquad (*)$$

$$\leq \max_{w_{[0,t_K]}} \left\{ |x(t_K)|^2_{\bar{Z}^+} + \sum_{k=0}^{K-1} F_k(x,w) \right\},$$

where the inequality follows because $\bar{Z}^+ > 0$. We now use dynamic programming to solve this problem. During the last sampling interval, we have the optimization problem:

$$\max_{w_{[t_{K-1}, t_K)}} |x(t_K)|^2_{\bar{Z}^+} + \int_{t_{K-1}}^{t_K} (|x|^2_Q + |B'\bar{Z}^+\Phi(t,t_k)x(t_k)|^2 - \gamma^2|w|^2)dt.$$

We now claim that whenever $\gamma > \hat{\gamma}^{SD}_\infty$, the maximum above exists, and the maximizing w is given by

$$w(t) = \gamma^{-2}D'\bar{Z}^+\Phi(t,t_{K-1})x(t_{K-1}).$$

Toward proving this claim, it is convenient to introduce a zero-sum soft-constrained differential game of the type defined by (4.1) and (4.2b), but with the interval being $[t_{K-1}, t_K)$. In this game, we take Players 1 and 2 to have open-loop information (that is they only have access to the value of $x(t_{K-1})$). This is consistent with the nature of the information pattern for Player 1, and for Player 2 we know that the precise nature of his information pattern is irrelevant since (as we shall shortly see) the upper value of the game is bounded whenever $\gamma > \hat{\gamma}^{SD}_\infty$. Now, in view of Theorem 4.1, a unique saddle-point solution to this open-loop differential game exists and is given by (4.6a)-(4.6b), where both (4.3) and (4.5) have as terminal (boundary) conditions $\bar{Z}^+$, replacing Q_f. With these boundary conditions, the RDE (4.3) is precisely (4.36) (on the appropriate sampling interval), and the solution of (4.5) is $\bar{Z}^+$ for all $t \in [t_{K-1}, t_K]$. Furthermore, the saddle-point strategies (4.6a) and (4.6b) are exactly (4.34a) and (4.34b), respectively. Hence, the game has a value, given by (in view of (4.8)) $x(t_{K-1})'\bar{Z}^+x(t_{K-1})$, which directly implies that

$$\max_{w_{[0,t_K]}} \{|x(t_K)|^2_{\bar{Z}^+} + \sum_{k=0}^{K-1} F_k(x,w)\}$$
$$= \max_{w_{[0,t_{K-1}]}} \{|x(t_{K-1})|^2_{\bar{Z}^+} + \sum_{k=0}^{K-2} F_k(x,w)\}.$$

On every sampling interval, we now have identical problems, and hence (inductively) the maximum above is equal to

$$x_0' \bar{Z}^+ x_0 ,$$

which provides a uniform (upper) bound for the finite-horizon maximization problem (∗). Since we already know that this value is the limiting *value* of the sequence of finite-horizon games, it follows that the control μ_∞^{SD} attains it, and the steady state controller (4.34b), or even (4.28b) (if disturbance is allowed to depend on the current value of the state), maximizes $J_\gamma^\infty(\mu_\infty^{\text{SD}}, w)$.

Finally, part (iv) follows from the boundedness of the upper value, which is $x_0' \bar{Z}^+ x_0$. If the hybrid system were not stable, then the upper value could be driven to $+\infty$ by choosing $w \equiv 0$, since $Q > 0$. ◇

c) Delayed perfect state measurements :

The infinite-horizon counterpart of Theorem 4.4 can be obtained by following arguments quite similar to those that led to Theorem 4.9. The counterpart of (4.36) here is the time-invariant version of the RDE (4.21):

$$\dot{S}_\theta + A'S_\theta + S_\theta A + Q + \frac{1}{\gamma^2} S_\theta DD' S_\theta = 0; \quad S_\theta(\theta) = \bar{Z}^+; \quad 0 \leq t < \theta, \quad (4.38)$$

which is now a single RDE, identical to (4.36), and defined on the time interval $[0, \theta]$. For boundedness of an upper value to the differential game, we now have to require, as in the sampled-data case, (in addition to the condition of Theorem 4.8) that the RDE (4.38) have no conjugate point on the given interval. To make this condition more precise, we introduce the following counterpart of (4.37):

$\hat{\gamma}_\infty^{\theta D} := \inf\{\gamma > 0 : \text{ARE (4.29) admits a positive definite solution, and}$

$\text{RDE (4.38) does not have a conjugate point on the interval } [0, \theta]\}.$
(4.39)

Then, the following theorem is a replica of Theorem 4.9, for the delayed information case.

Theorem 4.10. *Consider the infinite-horizon time-invariant linear-quadratic differential game, where $Q_f = 0$, the pair (A, H) is observable, with $H'H = Q$, and the information structure is delayed-state. Let $\hat{\gamma}_\infty^{\theta D}$ be defined as in (4.39). Then,*

(i) *If $\gamma > \hat{\gamma}_\infty^{\theta D}$, the game has equal upper and lower values, with the common value being (4.31a).*

(ii) *If $\gamma > \hat{\gamma}_\infty^{\theta D}$, the controller $\mu_\infty^{\theta D}$ given by*

$$\mu_\infty^{\theta D}(\eta(t)) = -B'\bar{Z}^+ \eta(t), \quad t \geq 0 \tag{4.40}$$

where $\eta(\cdot)$ is given by (the time-invariant version of) (4.20), attains the finite value, that is

$$\sup_{w \in \mathcal{H}_w} J_\gamma^\infty(\mu_\infty^{\theta D}, w) = x_0' \bar{Z}^+ x_0.$$

(iii) *If $\gamma < \hat{\gamma}_\infty^{\theta D}$, the upper value of the differential game is unbounded.*

(iv) *If $\gamma > \hat{\gamma}_\infty^{\theta D}$, Q is positive definite, and the controller (4.40) is used for u, the resulting closed-loop system described by a delayed differential equation becomes bounded input–bounded output stable.* ◇

4.4.2 The disturbance attenuation problem

The solution to the (infinite-horizon) H^∞-optimal control problem under the three information patterns of this chapter, can now readily be obtained from Theorems 4.8 – 4.10, by essentially following the arguments that led to the results of Section 4.3 (in the finite horizon) from the soft-constrained counterparts first developed in Section 4.2. We present these results below, without repeating some of the common arguments for their formal justification.

a) **Closed-loop perfect state measurements :**

Let us introduce the scalar

$$\hat{\gamma}_\infty^{\text{CL}} := \inf\{\gamma > 0 : \text{ARE (4.29) has a positive definite solution}\}. \quad (4.41)$$

Note that it is quite possible that the set above whose infimum determines $\hat{\gamma}_\infty^{\text{CL}}$ is empty. To avoid such pathological cases, we now assume (in addition to the observability condition of Theorem 4.8) that the pair (A, B) is *stabilizable*. This ensures, by a continuity argument applied to (4.29) at $\gamma = \infty$, that the set in (4.41) is nonempty, and hence $\hat{\gamma}_\infty^{\text{CL}}$ is finite.[4] Then, because of the established relationship between the soft-constrained differential game and the disturbance attenuation problem, the following result follows from Theorem 4.8.

Theorem 4.11. *For the continuous-time infinite-horizon disturbance attenuation problem (with $x_0 = 0$), let the condition of Theorem 4.8 on observability be satisfied, and the pair (A, B) be stabilizable. Then*

$$\gamma_\infty^* := \inf_{\mu \in \mathcal{M}_{\text{CLPS}}} \ll T_\mu \gg = \hat{\gamma}_\infty^{\text{CL}},$$

which is finite and is given by (4.41). Moreover, given any $\epsilon > 0$, we have the bound

$$\ll T_{\mu_{\gamma_\epsilon}^\infty} \gg \le \gamma_\epsilon := \hat{\gamma}_\infty^{\text{CL}} + \epsilon$$

where

$$\mu_\gamma^\infty(x) = -B' \bar{Z}_\gamma^+ x(t), \quad \gamma > \hat{\gamma}_\infty^{\text{CL}}, \quad (4.42)$$

with $\bar{Z}_\gamma^+$ being the unique minimal positive definite solution of (4.29). Furthermore, for any $\epsilon > 0$, $\mu_{\gamma_\epsilon}^\infty$ leads to a bounded input–bounded output stable system. ◊

---

[4] For a similar argument used in the discrete time, see the proof of Theorem 3.8. Note that with $\gamma = \infty$, the ARE (4.29) becomes the standard ARE that arises in linear-quadratic regulator theory ([56], [2]), which is known to admit a unique positive definite solution under the given conditions.

Remark 4.4. Again, as in the case of Theorem 4.8, we are not claiming that there exists an optimal controller (corresponding to $\gamma = \hat{\gamma}_\infty^{\text{CL}}$), because the ARE (4.29) may not admit any bounded solution for this limiting value of γ; the example of Section 4.4.3 will further illustrate this point. ◇

b) **Sampled-data perfect state information pattern :**

Under the sampled-data information pattern, it follows from the definition of $\hat{\gamma}_\infty^{\text{SD}}$, as given by (4.37), that

$$\hat{\gamma}_\infty^{\text{SD}} \geq \hat{\gamma}_\infty^{\text{CL}}$$

which is of course a natural relationship because the CLPS information pattern includes the SDPS pattern. In view of this relationship, the following theorem now follows from Theorem 4.9.

Theorem 4.12. *For the infinite-horizon disturbance attenuation problem with sampled state measurements, let the pair* $(A, Q^{1/2})$ *be observable, and* (A, B) *be stabilizable. Then,*

i) *The limiting (as $t_f \to \infty$) minimax attenuation level $\gamma_\infty^{\text{SD}}$, as defined by the infinite-horizon version of (1.2a) for the SDPS information, is equal to $\hat{\gamma}_\infty^{\text{SD}}$ given by (4.37), which is finite.*

ii) *Given any $\epsilon > 0$, and a corresponding $\gamma_{\infty,\epsilon} := \hat{\gamma}_\infty^{\text{SD}} + \epsilon$, the controller*

$$\mu_{\gamma_{\infty,\epsilon}}^\infty(t, x(t_k)) = -B' \bar{Z}_{\gamma_{\infty,\epsilon}}^+ \exp\{F_{\gamma_{\infty,\epsilon}}(t - t_k)\} x(t_k), \quad t_k \leq t < t_{k+1},$$
(4.43)

where

$$F_{\gamma_{\infty,\epsilon}} = A - \left(BB' - \frac{1}{\gamma_{\infty,\epsilon}^2} DD'\right) \bar{Z}_{\gamma_{\infty,\epsilon}}^+,$$
(4.44)

achieves an asymptotic performance no worse than $\gamma_{\infty,\epsilon}$. In other words,

$$\|T_{\mu_{\gamma_{\infty,\epsilon}}^\infty}(w)\| \leq \gamma_{\infty,\epsilon} \|w\|, \quad \text{for all } w \in \mathcal{H}_w.$$

iii) If $Q > 0$, the controller given by (4.43) leads to a bounded input–bounded output stable system.

iv) The following limiting controller exists:
$$\mu_{\gamma_\infty^{SD}}^\infty = \lim_{\epsilon \downarrow 0} \mu_{\gamma_{\infty,\epsilon}}^\infty.$$
◇

Remark 4.5. If the choice of the sampling times is also part of the design, then for the finite-horizon version and under the constraint of a fixed total number of sampled measurements in the given time interval, uniform sampling ($i.e., t_{k+1} - t_k =$ constant) will not necessarily be optimal, as we have seen earlier. For the time-invariant infinite-horizon version, however, uniform sampling will be overall optimal (under say an average frequency constraint), since the asymptotic performance is determined by the length of the longest sampling interval. ◇

c) Delayed state measurements :

Now, finally, under the delayed state measurement information pattern, Theorem 4.10 leads to the following solution for the disturbance attenuation problem.

Theorem 4.13. *For the infinite-horizon disturbance attenuation problem with θ-delayed perfect state measurements, let the pair $(A, Q^{1/2})$ be observable, and the pair (A, B) be stabilizable. Then,*

i) *The limiting (as $t_f \to \infty$) minimax attenuation level $\gamma_\infty^{\theta D}$, as defined by the infinite-horizon version of (1.2a) for the DPS information, is equal to $\hat{\gamma}_\infty^{\theta D}$ given by (4.39), which is finite.*

ii) *Given any $\epsilon > 0$, and a corresponding $\gamma_{\infty,\epsilon}^\circ := \hat{\gamma}_\infty^{\theta D} + \epsilon$, the controller*
$$\mu_{\gamma_{\infty,\epsilon}^\circ}^\infty(t, x(t_k)) = -B' \bar{Z}_{\gamma_{\infty,\epsilon}^\circ}^+ \eta(t), \quad t \geq \theta, \tag{4.45}$$

where $\eta(\cdot)$ is generated by (the time-invariant version of) (4.20), achieves an asymptotic performance no worse than $\gamma^\circ_{\infty,\epsilon}$. In other words,

$$\|T_{\mu^\infty_{\gamma^\circ_{\infty,\epsilon}}}(w)\| \leq \gamma^\circ_{\infty,\epsilon}\|w\|, \quad \text{for all } w \in \mathcal{H}_w. \tag{4.46}$$

iii) If $Q > 0$, the controller given by (4.45) leads to a bounded input–bounded output stable system.

iv) The following limiting controller exists:

$$\mu^\infty_{\gamma^\circ_\infty D} = \lim_{\epsilon \downarrow 0} \mu^\infty_{\gamma^\circ_{\infty,\epsilon}}. \tag{4.47}$$

Remark 4.6. Note that, comparing the conditions of Theorems 4.12 and 4.13, we arrive at the rather interesting observation that the optimal attenuation levels under the sampled-data and delayed information patterns (γ^{SD}_∞ and γ°_∞) are equal, whenever the length of the longest sampling interval (i.e., $\bar{t}_s$) and the length of delay (i.e., θ) are equal. This is so in spite of the fact that the controllers used are (structurally) different. ◇

4.4.3 Illustrative example (continued)

We now continue with the illustrative example of Section 4.3.3, this time for the infinite-horizon case.

The minimax attenuation level under CLPS information is $\gamma^*_\infty = 1$, and for $\gamma > \gamma^*_\infty$ the unique positive solution of the ARE (4.29) is

$$\bar{Z}^+_\gamma = \gamma/\sqrt{\gamma^2 - 1}, \quad \gamma > 1.$$

Note that an optimal controller does not exist under the CLPS information pattern, since $\bar{Z}^+_\gamma$ becomes unbounded as $\gamma \downarrow 1$.

For the sampled-data case, under uniform sampling (which is "optimal" in the infinite-horizon case, in view of Remark 4.5), and with a sampling time of t_s, the conjugate-point condition of

$$\dot{S} + 1 + \frac{1}{\gamma^2}S^2 = 0; \quad S(t_s) = \bar{Z}_\gamma^+$$

(i.e., requiring $S(t)$ to be bounded for $t \in [0, t_s]$), leads to the inequality

$$t_s + \gamma \arctan[1/\sqrt{\gamma^2 - 1}] < \frac{\pi}{2}\gamma. \qquad (**)$$

Hence, for a given t_s, the minimax attenuation level is the smallest positive value of γ that solves $(**)$ as an equality. Numerical experimentation has in fact shown that the solution to $(**)$ is actually unique, so that we do not have to search for the smallest positive root. Furthermore, this unique solution converges to γ_∞^* (from above) as $t_s \downarrow 0$. We give below the unique solution of $(**)$ for three different choices of t_s:

$$t_s = \pi \Rightarrow \gamma_\infty^{SD} \approx 2.653; \quad t_s = \frac{\pi}{2} \Rightarrow \gamma_\infty^{SD} \approx 1.682; \quad t_s = \frac{\pi}{4} \Rightarrow \gamma_\infty^{SD} \approx 1.241.$$

Note that γ_∞^{SD} is a monotonically increasing function of t_s, which is in fact a general property of the solution (in infinite-horizon problems) as already mentioned. Note also that since $\gamma_\infty^{SD} > 1 = \gamma_\infty^*$ in each case, the corresponding optimal controllers are asymptotically well-defined as $\gamma \downarrow \gamma_\infty^{SD}$, with the limits given (using (4.45)) by

$$\mu_\gamma^\infty(t, x(t_k)) = -\exp\left\{-\sqrt{\frac{\gamma-1}{\gamma+1}}(t - t_k)\right\} x(t_k), \quad t_k \leq t < t_k + t_s,$$

where $\gamma = \gamma_\infty^{SD}$.

4.5 More General Classes of Problems

Heretofore we have presented a complete theory for the special class of systems described by (4.1)-(4.2), where the controlled output is the state and the initial state is known (actually *zero* in the disturbance attenuation problem). This section, which parallels the development of Section 3.5, shows how this theory readily applies (with minor modifications) to the general class of continuous-time systems formulated in Section 1.3, and when the initial state x_0 is possibly an unknown quantity.

4.5.1 A more general cost structure

If we have the more general cost structure (1.10a) with z given by (1.9b), the results of Sections 4.2.2 and 4.3.1 readily apply, provided that we make the transformations

$$u(t) \to \widetilde{u}(t) := R^{\frac{1}{2}} \left[u(t) + R^{-1} G' H x(t) \right] \tag{4.48}$$

$$A(t) \to \widetilde{A}(t) := A(t) - B R^{-1} G' H \tag{4.49a}$$

$$B(t) \to \widetilde{B}(t) := B(t) R^{-\frac{1}{2}} \tag{4.49b}$$

$$Q(t) \to \widetilde{Q}(t) := Q - H' G R^{-1} G' H \geq 0. \tag{4.49c}$$

where $R := G'G, \quad Q := H'H$.

The fundamental (closed-loop) Riccati equation is still given by (4.5), with A, Q and B replaced by $\widetilde{A}$, $\widetilde{Q}$ and $\widetilde{B}$, respectively. Denoting its solution by $\widetilde{Z}$, the strongly time-consistent feedback controller (4.11a) is now expressed as

$$\mu_\gamma^*(t, x(t)) = -R^{-1} \left[B' \widetilde{Z}_\gamma(t) + G' H \right] x(t) \tag{4.50}$$

from which the optimal (or ϵ-optimal) H^∞-controller follows for $\gamma = \widetilde{\gamma}^{\text{CL}}$ (or $\gamma = \widetilde{\gamma}^{\text{CL}} + \epsilon, \epsilon > 0$), where $\widetilde{\gamma}^{\text{CL}}$ is defined by (4.10) with the Z- Riccati equation (4.5) replaced by the $\widetilde{Z}$- Riccati equation introduced above.

If z given by (1.9b) has also a linear term in w, then a transformation of the type (4.48) first applied to w (now, as a function of u and x) would bring it to the form of the problem considered above, and hence the previous theory can again be used. We should note, however, that this procedure (of eliminating the cross terms between w and u by expressing w as a function of u, in addition to x) leads to the *upper value* of the associated soft-constrained game (which is really what is sought) which may be different from its *lower value*.

4.5.2 Unknown nonzero initial state

If the initial state of the system is completely unknown (instead of being *zero*), one possible approach is to treat it also as a disturbance, as already discussed for the discrete-time problem in Section 3.5.2. One way of accommodating this generalization in the formulation of Chapter 1 (Section 1.2) would be to replace (1.2b) by ([51], [82])

$$\ll T_\mu \gg := \sup_{w \in \mathcal{H}_w,\ x_0 \in \mathbf{R}^n} \left\{ ||T_\mu(w)||_z \Big/ \{||w||_w^2 + x_0' Q_0 x_0\}^{1/2} \right\} \quad (4.51)$$

where $x(0) = x_0$, and Q_0 is an appropriately chosen nonnegative-definite weighting matrix. Now, using Theorem 4.2, and the argument that led to Theorem 4.5, we arrive at the relationship

$$\max_{w,x_0}\{||T_{\mu_\gamma^*}(w)||_z^2 - \gamma^2(||w||_w^2 + x_0' Q_0 x_0)\} = \max_{x_0}\{x_0'(Z_\gamma(0) - \gamma^2 Q_0) x_0\} \quad (4.52)$$

where we again look for the smallest value of $\gamma > 0$ for which the right-hand side is bounded (actually *zero*). In view of this, the set whose *infimum* determines $\widehat{\gamma}^{\mathrm{CL}}$ in (4.10) is now the set of all positive scalars γ for which the conjugate-point condition of the RDE (4.5) is satisfied *and* the matrix inequality

$$\gamma^2 Q_0 - Z_\gamma(0) \geq 0 \quad (4.53)$$

holds. This new set will determine a new optimum performance $\gamma^* = \widehat{\gamma}^{\mathrm{CL}}$, from which again the natural counterpart of Theorem 4.5 will follow. We should note, however, that it is quite possible (depending on the choice of Q_0) that the condition (4.53) will not impose any additional restrictions on γ^*, in particular if $Z_\gamma(\cdot)$ has finite escape in the interior (rather than on the boundary) of the time interval.

For the infinite-time horizon, one can analogously introduce the counterpart of (4.41), thereby arriving at the counterpart of Theorem 4.8 for the problem with unknown initial state. The same extension also applies to the sampled-data and delayed-state information cases.

4.6 Nonlinear Systems and Nonquadratic Performance Indices

We show in this section how the general methodology of this chapter can be used, along with Theorem 2.6, to obtain a disturbance attenuation bound for nonlinear systems with nonquadratic cost functions. Toward this end, let (4.1a) and (4.2a) be replaced by (2.13a) and (2.14), respectively, which we rewrite here for convenience (with zero initial state):

$$\dot{x} = f(t; x(t), u(t), w(t)), \quad x(0) = 0, \quad t \geq 0 \quad (4.54)$$

$$L(u, w) = q(x(t_f)) + \int_0^{t_f} g(t; x(t), u(t), w(t))dt. \quad (4.55a)$$

The associated soft-constrained game has the performance index

$$L_\gamma(u, w) = L(u, w) - \gamma^2 \|w\|^2. \quad (4.55b)$$

Consider the *Isaacs'* equation associated with this differential game, which is given by (2.16) with g replaced by g_γ, where the latter is defined by

$$g_\gamma(t; x, u, w) := g(t; x, u, w) - \gamma^2 |w|^2. \quad (4.55c)$$

Denote the solution of *Isaacs'* equation (i.e., the value function) by $V_\gamma(t; x)$, where have indicated explicitly the dependence on the positive parameter γ. Let the minimizing control in (2.16) (again with g_γ replacing g) be

$$u^*(t) = \mu_\gamma^*(t; x(t)), \quad 0 \leq t \leq t_f. \quad (4.56)$$

Note that this is necessarily a feedback controller, as discussed in Section 2.3.

Now, finally, introduce the following counterpart of (4.10):

$$\widehat{\gamma}^{\mathrm{NCL}} := \inf\{\gamma \in \Gamma^{\mathrm{NCL}}\} \quad (4.57a)$$

$$\Gamma^{\mathrm{NCL}} := \{\tilde{\gamma} > 0 : \forall \gamma \geq \tilde{\gamma}, \quad V_\gamma \text{ exists}\}. \quad (4.57b)$$

The following theorem, which is a weaker version of Theorem 4.5 for nonlinear systems, now follows from the discussion of Section 1.2, and in view of Theorem 2.6.

Theorem 4.14. *For the nonlinear/nonquadratic problem introduced in this section, let the value function of the soft-constrained game, $V_\gamma(t;x)$, satisfy the property*

$$V_\gamma(0;0) = 0, \quad \forall \gamma > \hat{\gamma}^{\text{NCL}}. \tag{4.58}$$

Then, for every $\gamma > \hat{\gamma}^{\text{NCL}}$, the feedback controller μ_γ^* ensures a disturbance attenuation bound γ, that is

$$\|T_{\mu_\gamma^*}(w)\|^2 := J(\mu_\gamma^*, w) \leq \gamma^2 \|w\|^2, \quad \forall w \in \mathcal{H}_w. \tag{4.59}$$

◇

Note that Theorem 4.14 is weaker than Theorem 4.5, because it does not claim that $\hat{\gamma}^{\text{NCL}}$ is the optimal (minimax) disturbance attenuation level. For a stronger result one has to bring in further structure on f and g, and derive (for nonlinear systems) the counterparts of the necessary and sufficient conditions of Appendix A.

4.7 Main Points of the Chapter

This chapter has presented the solution to the continuous-time disturbance attenuation (H^∞-optimal control) problem under several perfect state information patterns. When the controller has access to the current value of the state, the optimum attenuation level is determined by the nonexistence of a conjugate point to a particular Riccati differential equation in the finite horizon (see Theorem 4.5), and by the nonexistence of a positive definite solution to an algebraic Riccati equation in the infinite horizon case (see Theorem 4.11). In contradistinction with the discrete-time case, however, a corresponding optimal controller may not exist. There exist suboptimal linear feedback controllers ((4.23c) or (4.42)) ensuring attenuation levels arbitrarily close to the optimum ones.

Under sampled-data measurements, a set of additional RDE's (4.15), whose boundary conditions are determined by the solution of the CLPS

RDE (4.5), play a key role in the determination of the optimum attenuation level (see Theorem 4.6). A controller that achieves a performance level arbitrarily close to the optimum is again a linear feedback rule (4.26a), that uses the most recent sampled value of the state. For the infinite-horizon version, the optimum attenuation level is determined by one ARE and one RDE, with the latter determined on the longest sampling interval (see Theorem 4.12). The same conditions apply to the infinite-horizon delayed state measurement problem as well, even though the near-optimal controllers in this case have completely different structures (see Theorem 4.13; and Theorem 4.7 for the finite horizon case). The chapter has also discussed extensions to problems with (i) performance indices containing cross terms between state and control, (ii) nonzero initial states, and (iii) nonlinear dynamics/nonquadratic cost functions.

As mentioned earlier, the H^∞-optimal control problem under the CLPS information pattern was essentially solved in [65], which, however, does not contain a complete solution (in terms of both necessary and sufficient conditions) to the finite horizon case. The sampled-data solution, as given here, was first presented in [14] and was further discussed in [8] and [12], which also contain results on the DPS information case. The continuous-time H^∞-optimal control problem under CLPS information and using the state space approach has been the subject of several (other) papers and theses, among which are [53], [59], [71], [86], [80], and [51]. A preliminary study of the nonlinear problem, using game-theoretic techniques, can be found in [45].

Chapter 5

The Continuous-Time Problem With Imperfect State Measurements

5.1 Formulation of the Problem

We now turn to the class of continuous-time problems originally formulated in Section 1.3, where the state variable is no longer available to the controller, but only a disturbance-corrupted linear output is. The system is therefore described by the following equations:

$$\dot{x}(t) = A(t)x(t) + B(t)u(t) + D(t)w(t), \quad x(0) = x_0 \tag{5.1}$$

$$y(t) = C(t)x(t) + E(t)w(t). \tag{5.2}$$

We have not included an explicit dependence on u in the output equation (5.2), as the controller knows u and can always subtract out such a term from y. The regularity assumptions on the matrices $C(\cdot)$ and $D(\cdot)$ are as on A, B, D in the previous chapter: piecewise-continuous and bounded.

To emphasize duality, it will be helpful to write the performance index in terms of a second output (as in (1.9)-(1.10))

$$z(t) = H(t)x(t) + G(t)u(t), \tag{5.3}$$

where we first take

$$H'(t)H(t) =: Q(t), \tag{5.4a}$$

$$G'(t)G(t) = I, \tag{5.4b}$$

$$H'(t)G(t) = 0, \qquad (5.4c)$$

so that we may write the performance index, to be minimized, as

$$J(u,w) = |x(t_f)|^2_{Q_f} + \|x\|^2_Q + \|u\|^2 \equiv |x(t_f)|^2_{Q_f} + \|z\|^2. \qquad (5.5)$$

For ease of reference, we reproduce here the solution, given in Section 4.3.1, of the full state min-max design problem, for a given attenuation level γ:

$$\dot{Z} + ZA + A'Z - Z(BB' - \gamma^{-2}DD')Z + Q = 0, \quad Z(t_f) = Q_f \qquad (5.6)$$

$$u^*(t) = \mu^*(t; x(t)) = -B'(t)Z(t)x(t). \qquad (5.7a)$$

As shown in Section 4.2 (Theorem 4.2), for a related "soft-constrained" differential game, this controller is in saddle-point equilibrium with the "worst-case" disturbance

$$w^*(t) = \nu^*(t; x(t)) = \gamma^{-2}D'(t)Z(t)x(t). \qquad (5.7b)$$

Notice that (5.4b) implies that G is injective. Dually, we shall assume that E *is surjective*, and we shall let

$$E(t)E'(t) =: N(t), \qquad (5.8a)$$

where $N(t)$ is invertible.[1] To keep matters simple, and symmetrically to (5.4c) above, we shall, in the main development, assume that

$$D(t)E'(t) = 0. \qquad (5.8b)$$

However, in Section 5.4, we will provide a more general solution to the problem, to cover also the cases where the restrictions (5.4c) and (5.8b) are not imposed.

The admissible controllers,

$$u = \mu(y) \qquad (5.9)$$

---

[1] As a matter of fact, it would be possible to absorb $N^{-\frac{1}{2}}$ into D and E, and take $EE' = I$. We choose not to do so because it is often interesting, in examples and practical applications, to vary N, which is a measure of the measurement noise intensity.

the set of which will simply be denoted by $\mathcal{M}$, are controllers which are *causal*, and under which the triple (5.1), (5.2), (5.9) has a unique solution for every x_0, and every $w(\cdot) \in L^2([0,t_f], I\!R^n) = \mathcal{W}$. (In the infinite-horizon case, we shall impose the further condition that (5.9) stabilizes the system, including the compensator).

We shall consider, as the standard problem, the case where x_0 is part of the unknown disturbance, and will show on our way how the solution is to be modified if, to the contrary, x_0 is assumed to be known, and equal to zero. To simplify the notation, we shall let

$$(x_0, w) =: \omega \in \Omega := I\!R^n \times \mathcal{W}. \tag{5.10}$$

We also introduce the extended performance index

$$J_\gamma(u,w) = |x(t_f)|^2_{Q_f} + \|x\|^2_Q + \|u\|^2 - \gamma^2[\|w\|^2 + |x_0|^2_{Q_0}] \tag{5.11}$$

where Q_0 is a weighting matrix, taken to be *positive definite*.

In view of the discussion of Section 1.2, the disturbance attenuation problem to be solved is the following:

Problem $\mathcal{P}_\gamma$. Determine necessary and sufficient conditions on γ such that the quantity

$$\inf_{\mu \in \mathcal{M}} \sup_{\omega \in \Omega} J_\gamma(\mu, \omega)$$

is finite (which implies that then it will be zero), and for each such γ find a controller μ (or a family of controllers) that achieves the minimum. The infimum of all γ's that satisfy these conditions will be denoted by γ^*. ◇

We present the solution to this basic problem, in the finite horizon, in Section 5.2. Section 5.3 deals with the sampled-data case, and Section 5.4 studies the infinite horizon case, also known as the *four block problem*. Section 5.5 extends these to more general classes of problems with cross terms in the cost function, and with delayed measurements; it also discusses some extensions to nonlinear systems. The chapter concludes with Section 5.6 which summarizes the main results.

5.2 A Certainty Equivalence Principle, and Its Application to the Basic Problem $\mathcal{P}_\gamma$

To simplify the notation in the derivation of the next result, we shall rewrite (5.1)-(5.2) as

$$\dot{x} = f(t; x, u, w), \quad x(0) = x_0, \tag{5.12a}$$

$$y = h(t; x, w) \tag{5.12b}$$

and (5.11) as

$$J_\gamma(u, w) = q(x(t_f)) + \int_0^{t_f} g(t; x(t), u(t), w(t))dt + N(x_0). \tag{5.12c}$$

For a given pair of control and measurement trajectories, $\bar{u} = \bar{u}_{[0,t_f]}$ and $\bar{y} = \bar{y}_{[0,t_f]}$, we shall introduce a family of constrained optimization problems $Q^\tau(\bar{u}, \bar{y})$, indexed by $\tau \in [0, t_f]$, called "auxiliary problems", in the following way. Let

$$\Omega_\tau(\bar{u}, \bar{y}) = \{\omega \in \Omega \mid y(t) = \bar{y}(t), \quad \forall t \in [0, \tau]\}. \tag{5.13}$$

Here, of course, $y(t)$ is generated by $\bar{u}$ and ω through (5.1)-(5.2). Notice that Ω_τ depends only on the restrictions of $\bar{u}$ and $\bar{y}$ to the subinterval $[0, \tau]$, that is on $\bar{u}^\tau := \bar{u}_{[0,\tau]}$ and $\bar{y}^\tau := \bar{y}_{[0,\tau]}$. Also, the property $\omega \in \Omega_\tau$ depends only on $\omega^\tau := (x_0, w^\tau) = (x_0, w_{[0,\tau]})$, and not on $w_{(\tau,t_f]}$, so that we shall also write, when convenient, $\omega^\tau \in \Omega_\tau^\tau(\bar{u}, \bar{y})$, or equivalently $\omega^\tau \in \Omega_\tau^\tau(\bar{u}^\tau, \bar{y}^\tau)$.

Let $\bar{u}$ and $\bar{y}$ be fixed. Obviously, if $\tau' > \tau$, $\Omega_{\tau'} \subset \Omega_\tau$. But also, because E is assumed to be surjective, any $\omega^\tau \in \Omega_\tau^\tau$ can be extended to an $\omega^{\tau'} \in \Omega_{\tau'}^{\tau'}$. Therefore, for a given pair $(\bar{u}, \bar{y})$, the set $\Omega_{\tau'}^\tau$, of the restrictions of the elements of $\Omega_{\tau'}$ to $[0, \tau]$, is Ω_τ^τ:

$$\forall \bar{u}, \bar{y}, \quad \forall \tau' > \tau, \quad \Omega_{\tau'}^\tau(\bar{u}, \bar{y}) = \Omega_\tau^\tau(\bar{u}, \bar{y}),$$

so that, finally,

$$\forall u \in \mathcal{U}, \quad \forall y \in \mathcal{Y}, \quad \forall \tau \in [0, t_f], \quad \Omega_\tau^\tau(u, y) = \Omega_{t_f}^\tau(u, y). \tag{5.14}$$

Consider also the soft-constrained zero-sum differential game with full state information, defined by (5.1) and (5.11) (or equivalently (5.12c)), but *without the $N(x_0)$ term*. Let (μ^*, ν^*) be its unique state-feedback strongly time consistent saddle-point solution (given by (5.7)), and let $V(t;x)$ $(=|x|^2_{Z(t)})$ be its value function. Finally introduce the performance index

$$G^\tau(u,\omega) = V(\tau;x(\tau)) + \int_0^\tau g(t;x,u,w)dt + N(x_0). \qquad (5.15)$$

We are now in a position to state the auxiliary problems:

Problem $Q^\tau(\bar{u}^\tau, \bar{y}^\tau)$

$$\max_{\omega^\tau \in \Omega^\tau_\tau(\bar{u},\bar{y})} G^\tau(\bar{u}^\tau, \omega^\tau). \qquad (5.16)$$

$\diamond$

This is a well-defined problem, and, as a matter of fact, $\Omega^\tau_\tau(\bar{u},\bar{y})$ depends only on $(\bar{u}^\tau, \bar{y}^\tau)$. However, it will be convenient to use the equivalent form

$$\max_{\omega \in \Omega_{t_f}(\bar{u},\bar{y})} G^\tau(\bar{u},\omega) \qquad (5.17)$$

since in fact G^τ depends only on $\bar{u}^\tau$ and ω^τ, and therefore (5.17) depends on $\Omega^\tau_{t_f}(\bar{u},\bar{y})$ which is $\Omega^\tau_\tau(\bar{u},\bar{y})$ in view of (5.14).

Remark 5.1. The maximization problem Q^τ defined by (5.16) has an obvious interpretation in terms of "worst case disturbance". We seek the worst overall disturbance compatible with the information available at time τ. Thus, x_0 and past values of w should agree with our own past control and past measurements, but future w's could be the worst possible, without any constraint imposed on them, i.e., chosen according to ν^*, leading to a least cost-to-go $V(\tau;x(\tau))$. $\diamond$

We now describe the derivation of a controller which will turn out to be an optimal "min sup" controller. Assume that for $(\bar{u}^\tau, \bar{y}^\tau)$ given, the problem Q^τ admits a unique maximum $\hat{\omega}^\tau$, and let $\hat{x}^\tau$ be the trajectory

Imperfect State Measurements: Continuous Time 119

generated by $(\bar{u}^\tau, \widehat{\omega}^\tau)$ up to time τ. Let

$$\widehat{u}(\tau) = \mu^*(\tau; \widehat{x}^\tau(\tau)). \tag{5.18}$$

This defines an implicit relation $\widehat{\mu}$, since $\widehat{x}^\tau(\tau)$ itself depends on $\widehat{u}^\tau$. We therefore have a relation $u(\tau) = \widehat{\mu}(u^\tau, y^\tau)$ whose fixed point is the control sought. In fact, because of the casuality inherent in the definition (5.18), this fixed point exists and is unique, and (5.18) turns out to provide an explicit description for $\widehat{\mu}(y)$.

Relation (5.18) says, in simple terms, that one should look, at each instant of time τ, for the worst possible disturbance $\widehat{\omega}^\tau$ compatible with the available information at that time, and with the corresponding worst possible state trajectory, and use the control which would agree with the optimal state feedback strategy if current state were the worst possible one. Thus this is indeed a *worst case certainty equivalence principle*, but not a separation principle since this worst case trajectory depends on the payoff index to be minimized.

Theorem 5.1. *If, for every* $y \in \mathcal{Y}$ *and every* $\tau \in [0, t_f]$, *the problem* $\mathcal{Q}^\tau(\widehat{\mu}(y), y)$ *has a unique maximum* $\widehat{\omega}^\tau$ *generating a state trajectory* $\widehat{x}^\tau$, *then (5.18) provides a min sup controller solution for problem* $\mathcal{P}_\gamma$, *and the minimax cost is given by*

$$\min_{\mu \in \mathcal{M}} \max_{\omega \in \Omega} J_\gamma = \max_{x_0 \in \mathbf{R}^n} \{V(0; x_0) + N(x_0)\} = V(0; \widehat{x}_0^0) + N(\widehat{x}_0^0).$$

Proof. Notice first that the maximum in the second expression above exists under the hypotheses of the theorem, and the notations in the third expression are consistent, since the problem $\mathcal{Q}^0$, assumed to have a unique maximum, is just that maximization problem.

Introduce the function

$$W^\tau(\bar{u}, \bar{y}) := \max_{\omega \in \Omega_{t_f}(\bar{u}, \bar{y})} G^\tau(\bar{u}, \omega). \tag{5.19}$$

Notice also that for fixed $\bar{u}$ and ω, one has

$$\frac{\partial}{\partial \tau} G^\tau(\bar{u}, \omega) = \frac{\partial V}{\partial \tau} + \frac{\partial V}{\partial x} f(\tau; x, \bar{u}, w) + g(\tau; x, \bar{u}, w),$$

with all functions on the right-hand side above being evaluated at $(\tau; x(\tau), \bar{u}(t), w(\tau))$. It now follows from a theorem due to Danskin (see Appendix B; Chapter 9), that, under the hypothesis that $\widehat{\omega}^\tau$ is unique, the function W^τ defined by (5.19) has a derivative in τ:

$$\frac{\partial W^\tau(\bar{u}, \bar{y})}{\partial \tau} = \frac{\partial V}{\partial \tau} + \frac{\partial V}{\partial x} f(\tau; \widehat{x}^\tau(\tau), \bar{u}(\tau), \widehat{w}^\tau(\tau)) + g(\tau; \widehat{x}^\tau(\tau), \bar{u}(\tau), \widehat{w}^\tau(\tau)).$$

Suppose that $\bar{u} = \widehat{\mu}(\bar{y})$, so that $\bar{u}(\tau)$ is chosen according to (5.18); then it follows from Isaacs' equation (2.16) that

$$\frac{\partial}{\partial \tau} W^\tau(\widehat{\mu}(\bar{y}), \bar{y}) \leq 0.$$

As a result, $W^\tau(\widehat{\mu}(\bar{y}), \bar{y})$ is a decreasing function of τ, and therefore

$$W^{t_f}(\widehat{\mu}(\bar{y}), \bar{y}) \leq W^0 = V(0; \widehat{x}_0^0) + N(\widehat{x}_0^0).$$

Moreover, since $V(t_f; x) = q(x)$, we have $G^{t_f} = J_\gamma$, and thus

$$\forall \omega \in \Omega, \quad J_\gamma(\widehat{\mu}(y), \omega) \leq W^{t_f}(\widehat{\mu}(y), y) \leq V(0; \widehat{x}_0^0) + N(\widehat{x}_0^0). \tag{5.20a}$$

On the other hand, for every given $u \in \mathcal{U}$, a possible disturbance is $x_0 = \widehat{x}_0^0$ along with a w that agrees with ν^*, and this leads to the bound

$$\forall u \in \mathcal{U} \quad J_\gamma(\widehat{x}_0^0; u, \nu^*) \geq V(0; \widehat{x}_0^0) + N(\widehat{x}_0^0). \tag{5.20b}$$

Comparing this with (5.20a) proves the theorem. ◇

Our derivation above has in fact established a stronger type of optimality than that stated in Theorem 5.1, namely

Corollary 5.1. *For a given τ, and any past u^τ, and under the hypotheses of Theorem 5.1, the use of (5.18) from τ onwards ensures*

$$J_\gamma \leq W^\tau(u^\tau, y^\tau)$$

and this is the best possible (smallest) guaranteed outcome. ◇

The above derivation holds for any optimal strategy μ of the minimizer. As we have discussed in Section 2.3, and also in Section 4.2.2, many such optimal strategies may exist which are all representations (cf. Definition 2.2) of μ^*. We shall exhibit such strategies of the form[2]

$$\mu_\psi(t;x,y) = \mu^*(t;x) + \gamma\psi(y - Cx)$$

where ψ is a causal operator from $\mathcal{Y}$ into U, with $\psi(0) = 0$. Placing this control into the classical "completion of the squares" argument (see (4.9) or (8.15)), we get

$$q(x(t_f)) + \int_\tau^{t_f} g(t;x,u,w)dt = V(\tau;x(\tau))$$

$$+\gamma^2 \int_\tau^{t_f} (-|w - \nu^*(t;x)|^2 + |\psi(y - Cx)|^2)dt.$$

Notice that $y - Cx = Ew$, and that $E\nu^*(t;x) = 0$ (from (5.8b)). Therefore, the integral on the right-hand side is stationary at $w = \nu^*(t;x)$, and maximum if, for instance,

$$\exists \alpha < 1 : \int_\tau^{t_f} |\psi(Ew)|^2 dt \leq \alpha \int_\tau^{t_f} |w|^2 dt.$$

In view of (5.8a), and for a fixed $w \in \mathcal{W}$ such that $Ew = \eta$, one easily sees that $|w|^2 \geq |\eta|^2_{N^{-1}}$, so that the above holds, ensuring that all games from τ to t_f have a saddle point if

$$\forall \tau \epsilon [0, t_f], \quad \exists \alpha : \forall \eta(\cdot), \int_\tau^{t_f} |\psi(\eta(t))|^2 dt \leq \alpha \int_\tau^{t_f} |\eta(t)|^2_{N^{-1}} dt. \quad (5.21)$$

Applying Theorem 5.1 with μ_ψ, and noticing that by construction $\widehat{y}^\tau = y^\tau$, we get:

---

[2] This parallels the analysis of Section 4.2.2 that led to (4.12b) in the CLPS information case.

Corollary 5.2. *A family of optimal controllers for problem* $\mathcal{P}_\gamma$ *is given by*

$$\widehat{\mu}_\psi(t;y) = \mu^*(t;\widehat{x}^{t_f}(t)) + \gamma\psi(y - C\widehat{x}^{t_f})(t) \tag{5.22}$$

where ψ is any causal operator from $\mathcal{Y}$ into $\mathcal{U}$ such that (5.21) is satisfied, and the existence and unicity of the solution of the differential equation (5.1) is preserved. ◇

We may notice that (5.21) can be seen as imposing that ψ be a contraction from $L^2([\tau,t_f], \mathbb{R}^p_{N-1})$ into $L^2([\tau,t_f], \mathbb{R}^{m_1})$, for all τ.

The following theorem is important for the application of the above result to our problem.

Theorem 5.2. *If for all $(u,y) \in \mathcal{U} \times \mathcal{Y}$, there exists a t^* such that the problem $\mathcal{Q}^{t^*}$ fails to have a solution and exhibits an infinite supremum, then the problem $\mathcal{P}_\gamma$ has no solution, the supremum in ω being infinite for any admissible controller $\mu \in \mathcal{M}$.*

Proof. Let $\bar{\mu}$ be an admissible controller, $\bar{\omega} \in \Omega$ be a disturbance, $\bar{y}$ be the output generated under the control $\bar{\mu}$, and $\bar{u} = \bar{\mu}(\bar{y})$. For every real number a, there exists, by hypothesis, a disturbance $\widetilde{\omega} \in \Omega_{t_f}(\bar{u},\bar{y})$ such that $G^{t^*}(\bar{u},\widetilde{\omega}) > a$. By the definition of Ω_{t_f}, the pair $(\bar{u},\widetilde{\omega})$ generates the same output $\bar{y}$ as $(\bar{u},\bar{\omega})$, so that $\widetilde{\omega}$ together with $\bar{\mu}$ will generate $(\bar{u},\bar{y})$. Consider the disturbance ω given by

$$x_0 = \widetilde{x}_0,$$

$$w(t) = \begin{cases} \widetilde{w}(t) & \text{if } t \leq t^* \\ \nu^*(x(t)) & \text{if } t > t^*. \end{cases}$$

Because of the saddle-point property of ν^*, it will cause $J_\gamma(\mu,\omega) \geq G^{t^*}(\bar{u},\widetilde{\omega}) \geq a$, from which the result follows. ◇

Remark 5.2. The results obtained so far in this section do not depend on the special linear-quadratic form of problem $\mathcal{P}_\gamma$. They hold equally well for nonlinear/nonquadratic problems, of the type covered in Section 4.6, provided that the conditions for the application of Danskin's theorem, as developed in Appendix B,[3] apply. However, it is due to the linear character of the problem that we shall always have either Theorem 5.1 or Theorem 5.2 to apply. Moreover, as we shall explain below, Theorem 5.1 can hardly be used to find an optimal controller unless some very special structure is imposed on the problem. ◇

We now undertake to use Theorem 5.1 to give a practical solution to problem $\mathcal{P}_\gamma$. The first step is to solve the generic problem $Q^\tau(u,y)$. A simple way to do this is to set it in a Hilbert space setting. Let $\mathcal{U}^\tau$, $\mathcal{W}^\tau$, $\mathcal{Y}^\tau$, $\mathcal{Z}^\tau$ be the Hilbert spaces $L^2([0,\tau], I\!\!R^k)$, with appropriate k's, where the variables u^τ, w^τ, y^τ and z^τ live. The system equations (5.1), (5.2), and (5.5) together define a linear operator from $\mathcal{U}^\tau \times \mathcal{W}^\tau \times I\!\!R^n$ into $\mathcal{Y}^\tau \times \mathcal{Z}^\tau \times I\!\!R^n$:

$$y^\tau = \mathcal{A}^\tau u^\tau + \mathcal{B}^\tau w^\tau + \eta^\tau x_0 \qquad (5.23a)$$

$$z^\tau = \mathcal{C}^\tau u^\tau + \mathcal{D}^\tau w^\tau + \zeta^\tau x_0 \qquad (5.23b)$$

$$x(\tau) = \varphi^\tau u^\tau + \psi^\tau w^\tau + \Phi^\tau x_0. \qquad (5.23c)$$

In the development to follow, we shall suppress the superscript τ, for simplicity in exposition. First we shall need the following lemma.

Lemma 5.1. *The operator dual to (5.23):*

$$\alpha = \mathcal{A}^* p + \mathcal{C}^* q + \varphi^* \lambda_f$$

$$\beta = \mathcal{B}^* p + \mathcal{D}^* q + \psi^* \lambda_f$$

$$\lambda_0 = \eta^* p + \zeta^* q + \Phi^* \lambda_f$$

---
[3] In the nonlinear case, hypothesis H2 need to be replaced by weak continuity of the map $\Omega \mapsto \dot{G}$.

admits the following internal representation, in terms of the vector function $\lambda(t)$:

$$-\dot\lambda(t) = A'(t)\lambda(t) + C'(t)p(t) + H'(t)q(t), \quad \lambda(\tau) = \lambda_f,$$

$$\alpha(t) = B'(t)\lambda(t) + G'(t)q(t),$$

$$\beta(t) = D'(t)\lambda(t) + E'(t)p(t),$$

$$\lambda_0 = \lambda(0).$$

Proof. The proof simply follows from a substitution of $\dot x$ and $\dot\lambda$ in the identity

$$\lambda'(\tau)x(\tau) - \lambda'(0)x(0) \equiv \int_0^\tau (\dot\lambda' x + \lambda' \dot x)dt,$$

leading to

$$\lambda'_f x(\tau) + \int_0^\tau (p'y + q'z)dt = \lambda'_0 x_0 + \int_0^\tau (\alpha' u + \beta' w)dt,$$

which proves the claim. $\diamond$

Problem $\mathcal{Q}^\tau$ may now be stated in terms of the operators introduced in (5.23), recalling that $V(\tau; x(\tau)) = |x(\tau)|^2_{Z(\tau)}$:

$$\max_{x_0, w} \Big[|\varphi u + \psi w + \Phi x_0|^2_{Z(\tau)} + \|\mathcal{C}u + \mathcal{D}w + \zeta x_0\|^2$$

$$- \gamma^2(\|w\|^2 + |x_0|^2_{Q_0})\Big]$$

under the constraint

$$\mathcal{A}u + \mathcal{B}w + \eta x_0 = y.$$

Since we assume that E is surjective, so is the operator $[\mathcal{A}\ \mathcal{B}\ \eta]$, and we may use a Lagrange multiplier to obtain the necessary conditions. Let $p \in L^2([0, \tau], \mathbb{R}^p)$ be the corresponding Lagrange multiplier. Differentiating with respect to w and x_0 successively, we get

$$\mathcal{B}^* p + \mathcal{D}^*(\mathcal{C}u + \mathcal{D}w + \zeta x_0) + \psi^* Z(\tau)(\varphi u + \psi w + \Phi x_0) = \gamma^2 w, \quad (5.24a)$$

$$\eta^* p + \zeta^*(\mathcal{C}u + \mathcal{D}\omega + \zeta x_0) + \Phi^* Z(\tau)(\varphi u + \psi w + \Phi x_0) = \gamma^2 Q_0 x_0. \quad (5.24b)$$

(In the case x_0 is known to be zero, we do not have equation (5.24b)). Using Lemma 5.1, with $q = \hat{z}$ and $\lambda_f = Z(\tau)x(\tau)$, equations (5.24) may be rewritten as

$$w = \gamma^{-2}(D'\lambda + E'p), \qquad (5.25a)$$

$$x_0 = \gamma^{-2}Q_0^{-1}\lambda(0). \qquad (5.25b)$$

The Lagrange multiplier p is obtained by placing (5.25a) into the constraint. A few substitutions then lead to

$$\dot{\hat{x}}^\tau = A\hat{x}^\tau + \gamma^{-2}DD'\lambda + B\hat{u}^\tau, \quad \hat{x}^\tau(0) = \gamma^{-2}Q_0^{-1}\lambda(0), \qquad (5.26a)$$

$$-\dot{\lambda} = -\gamma^2(C'N^{-1}C - \gamma^{-2}Q)\hat{x}^\tau + A'\lambda + \gamma^2 C'N^{-1}y,$$
$$\lambda(\tau) = Z(\tau)\hat{x}^\tau(\tau). \qquad (5.26b)$$

(And of course, if $x_0 = 0$ is fixed, the initial condition in (5.26a), which is (5.25b), should be replaced by $\hat{x}^\tau(0) = 0$).

This two-point boundary value problem should be solved for every τ, and $\hat{x}^\tau(\tau)$ be used in the optimal state feedback strategy, to yield

$$\hat{u}(\tau) = -B'(\tau)Z(\tau)\hat{x}^\tau(\tau). \qquad (5.27)$$

Unless we can find recursive formulas for $\hat{x}^\tau(\tau)$, this is hardly feasible. Toward this end, following the classical device of transforming two-point boundary value problems to Cauchy problems, we introduce a related Riccati differential equation in $\Sigma(t)$:

$$\dot{\Sigma} = A\Sigma + \Sigma A' - \Sigma(C'N^{-1}C - \gamma^{-2}Q)\Sigma + DD'; \qquad \Sigma(0) = Q_0^{-1} \quad (5.28)$$

and, if it has a solution over $[0, \tau]$, we further introduce

$$\check{x}(t) := \hat{x}^\tau(t) - \gamma^{-2}\Sigma(t)\lambda(t). \qquad (5.29)$$

A direct calculation shows that $\check{x}$ satisfies the following differential equation

$$\dot{\check{x}} = (A + \gamma^{-2}\Sigma Q)\check{x} + \Sigma C'N^{-1}(y - C\check{x}) + B\hat{u}, \quad \check{x}(0) = 0. \qquad (5.30)$$

(In the case $x_0 = 0$, set $\Sigma(0) = 0$, the rest is unchanged.) We now notice that (5.28) and (5.30) are independent of τ, provided that we can express $\widehat{x}^\tau(\tau)$ in terms of $\check{x}(\tau)$ to substitute in (5.27) and place this $\widehat{u}$ in (5.30). The final condition in (5.26b) gives, together with (5.29):

$$\check{x}(\tau) = \widehat{x}^\tau(\tau) - \gamma^{-2}\Sigma(\tau)Z(\tau)\widehat{x}^\tau(\tau)$$

so that, if the inverse below exists, we have

$$\widehat{x}^\tau(\tau) = \left(I - \gamma^{-2}\Sigma(\tau)Z(\tau)\right)^{-1} \check{x}(\tau). \qquad (5.31)$$

By substituting this into (5.27), we obtain the optimal controller given by

$$\widehat{u}(t) = -B'(t)Z(t)\left[I - \gamma^{-2}\Sigma(t)Z(t)\right]^{-1} \check{x}(t) \qquad (5.32)$$

where $\check{x}(t)$ is given by (5.30), into which we substitute $\widehat{u}$ according to (5.32).

A direct calculation, differentiating (5.31) and substituting (5.7), (5.28) and (5.30) into it, yields the following alternate formulas, where we have used $\widehat{x}(t)$ for $\widehat{x}^\tau(t)$:

$$\dot{\widehat{x}} = \left[A - (BB' - \gamma^{-2}DD')Z\right]\widehat{x} + [I - \gamma^{-2}\Sigma Z]^{-1}\Sigma C' N^{-1}(y - C\widehat{x}), \qquad (5.33)$$

and then (5.27) reads

$$\widehat{u}(t) = -B'(t)Z(t)\widehat{x}(t) \qquad (5.34)$$

Remark 5.3. If $Q = 0$, (5.28) and (5.30) become identical to the equations that characterize the standard Kalman filter [1]. This relationship between Kalman filtering and minimax estimation will be further elaborated on in Chapter 7. ◊

Proposition 5.1. *If either x_0 is unknown ($Q_0 > 0$), or (A, D) is completely controllable over $[t, t_f]$, the matrix $\Sigma(t)$ is positive definite whenever it is defined.*

Proof. In the case when x_0 is unknown, $\Sigma(0) = Q_0^{-1}$ is positive definite, and the result then follows from Proposition 8.4 given in Appendix A. In the case when x_0 is known (and fixed), $\Sigma(0) = 0$, and the result follows by duality with the Riccati equation of the control problem. By duality, complete controllability of (A, D) translates into complete observability of (A', D'), which is sufficient to ensure that $\Sigma > 0$. ◇

Notice also that the concavity of the problem $\mathcal{Q}^0$, and hence the existence of an $\hat{x}_0^0$ (which will necessarily be zero), necessitates that $Z(0) - \gamma^2 Q_0 < 0$, which is equivalent to the condition that the eigenvalues of $Q_0^{-1} Z(0) = \Sigma(0) Z(0)$, which are real, be of modulus less than γ^2. As a consequence, the condition that $(I - \gamma^{-2} \Sigma(t) Z(t))$ be invertible for all $t \in [0, t_f]$, translates into a condition on the spectral radius of $\Sigma(t) Z(t)$:

$$\forall t \in [0, t_f], \quad \rho(\Sigma(t) Z(t)) < \gamma^2, \tag{5.35a}$$

or equivalently, since $\Sigma(t)$ is invertible,

$$\forall t \in [0, t_f], \quad Z(t) - \gamma^2 \Sigma^{-1}(t) < 0. \tag{5.35b}$$

We now state the main result of this section.

Theorem 5.3. *Consider the disturbance attenuation problem of this section with continuous imperfect state measurements (5.2), and let its optimum attenuation level be γ^*. For a given $\gamma > 0$, if the Riccati differential equations (5.6) and (5.28) have solutions over $[0, t_f]$, and if condition (5.35) (in either of its two equivalent forms) is satisfied, then necessarily $\gamma \geq \gamma^*$. For each such γ, there exists an optimal controller, given by (5.30) and (5.32), or equivalently by (5.33) and (5.34), from which a whole family of optimal controllers can be obtained as in (5.22). Conversely, if either (5.6) or (5.28) has a conjugate point in $[0, t_f]$, or if (5.35) fails to hold, then $\gamma \leq \gamma^*$, i.e., for any smaller γ (and possibly for the one considered) the supremum in problem $\mathcal{P}_\gamma$ is infinite for any admissible μ.*

Proof. In view of Theorems 5.1 and 5.2, what we have to do is relate the concavity of the problems $\mathcal{Q}^\tau$ to the existence of Z, Σ and to condition (5.35). These problems involve the maximization of a nonhomogeneous quadratic form under an affine constraint, or equivalently over an affine subspace. The concavity of this problem only depends on the concavity of the homogeneous part of the quadratic form over the linear subspace parallel to the affine set. We may therefore investigate only the case obtained by setting $u = 0$ (homogeneous part of G^τ) and $y = 0$ (linear subspace).

Let us assume that Z and Σ exist over $[0, t_f]$, and that (5.35) holds. Let[4]

$$W(t; x) := \gamma^2 |x|^2_{\Sigma^{-1}(t)}. \tag{5.36}$$

It is now a simple matter, using Lagrange multipliers or *completion of the squares* method, and parameterizing all w's that satisfy $Cx + Ew = 0$ by

$$w = -E'N^{-1}Cx + (I - E'N^{-1}E)\nu, \quad \nu \in \mathbb{R}^{m_2},$$

to see that:[5]

$$\max_{w \mid Cx+Ew=0} \left[\frac{\partial W}{\partial t} + \frac{\partial W}{\partial x} f(t; x, 0, w) + g(t; x, 0, w) \right] = 0 \tag{5.37}$$

and moreover, $W(0; x) = -N(x)$. Integrating along the trajectory corresponding to an arbitrary w, we get

$$W(\tau; x(\tau)) + N(x_0) + \int_0^\tau g(t; x, 0, w) dt \leq 0.$$

with an equality sign for the maximizing w^τ. Hence we have,

$$\forall \omega \in \Omega_\tau(0, 0), \ G^\tau(0, \omega) \leq |x(\tau)|^2_{Z(\tau)} - W(\tau; x(\tau)) = |x(\tau)|^2_{Z(\tau) - \gamma^2 \Sigma^{-1}(\tau)}, \tag{5.38}$$

with an equality sign for the maximizing ω. Now, the condition (5.35b) implies that this is a concave function of $x(\tau)$, and hence that $G^\tau(0, \omega) \leq 0$ over $\Omega_\tau(0, 0)$. Thus the conditions of Theorem 5.1 apply, and the first part of Theorem 5.3 is proved.

---

[4] This function W is *not* the same as the one used in the proof of Theorem 5.1.

[5] We use the same notation f, g and N as in the statement and proof of Theorem 5.1.

The maximizing w in (5.37) is of the form $w = Px$ (with $P = D'\Sigma^{-1} - E'N^{-1}C$), P bounded. Hence, under that disturbance, the map $x_0 \mapsto x(\tau)$ is surjective. Therefore, the above derivation shows that if, at some τ, condition (5.35) does not hold, G^τ can be made arbitrarily large over $\Omega_\tau(0,0)$. So we are only left with the task of proving that existence of Σ over $[0, t_f]$ is necessary, since the necessity of the existence of Z to have $\gamma \geq \gamma^*$ is, as already seen, a direct consequence of Theorem 8.5.

Take any $(\widehat{x}, \lambda)$ satisfying the differential equations in (5.26), independently of the boundary conditions. Computing $\dfrac{d}{dt}(\lambda' x)$ and integrating, we get

$$\gamma^2[\lambda'(\tau)\widehat{x}(\tau) - \lambda'(0)\widehat{x}(0)] + \int_0^\tau (|x|_Q^2 - \gamma^2|w|^2)dt = 0. \tag{5.39}$$

Introduce the square matrices $X(t)$ and $Y(t)$ defined by[6]

$$\dot{X} = AX + DD'Y, \quad X(0) = Q_0^{-1},$$

$$\dot{Y} = (C'N^{-1}C - \gamma^2 Q)X - A'Y, \quad Y(0) = I.$$

All solutions of (5.26) that satisfy the first boundary condition (at 0, in (5.26a)) are of the form $x(t) = X(t)\xi$, $\lambda(t) = \gamma^2 Y(t)\xi$, where $\xi \in \mathbb{R}^n$ is a constant. But also, as long as Y, which is invertible at 0, remains invertible for all t, we have $\Sigma = XY^{-1}$, as direct differentiation of this last expression shows. Then, X is invertible, since Σ is nonsingular by Proposition 5.1. Now, let τ^* be such that $Y(\tau^*)$ is singular, and let $\xi \neq 0$ be such that $Y(\tau^*)\xi = 0$. Then $X(\tau^*)\xi \neq 0$, because the vector $(x'\ \lambda)' = (\xi'X'\ \gamma^2\xi'Y')'$, solution of a linear differential equation not zero at $t = 0$, can never be zero. Using $\widehat{x}_0 = Q_0^{-1}\xi^*$, the admissible disturbance $w = D'Y\xi^* - E'N^{-1}CX\xi^*$ leads to $\widehat{x} = X(t)\xi^*$ with $\lambda = Y(t)\xi^*$. Application of (5.39) then yields

$$G^{\tau^*} = |x(\tau^*)|^2_{Z(\tau^*)} \geq 0.$$

Therefore, for any smaller γ, since w here is nonzero, $G^{\tau^*}(0, w)$ will be positive, implying that the (homogeneous) problem is not concave. This completes the proof of the theorem. ◇

---
[6] These are Caratheodory's canonical equations, as revived by Kalman.

Remark 5.4. We may now check that the same theorem with the same formulas, except that $\Sigma(0) = 0$, hold for the problem where x_0 is given to be zero, and (A, D) is completely controllable. The sufficiency part can be kept with the additional classical device of Tonelli to handle the singularity at 0; see [30] or [21]. (The trick is to note that in a neighborhood of 0, a trajectory $X(t)\xi$ is optimal for the problem with free initial state, and with initial cost given by $-\gamma^2(|x_0|^2 + 2\xi'x_0)$. Therefore, any other trajectory reaching the same $x(t)$ gives a smaller cost, but for trajectories through $x(0) = 0$, the above initial cost is zero, so the costs compared are the same as in our problem).

The necessity of the existence of Σ is unchanged, and the necessity of (5.35) will again derive from (5.38) if we can show that any $\hat{x}(\tau)$ can be reached by maximizing trajectories. The controllability hypothesis ensures that $\Sigma(t) > 0$, but then Y being invertible, so is $X = \Sigma Y$. Since all reachable $\hat{x}(\tau)$ are precisely the points generated by $X(\tau)\xi$, $\xi \in \mathbb{R}^n$, this completes the proof of the result for the case when $x_0 = 0$.

Notice also that although this is not the way the result was derived, replacing Q_0^{-1} by the zero matrix amounts to placing an infinite penalty on nonzero initial states. ◇

Proposition 5.2. *For the problem with $x_0 = 0$, let γ_0^* be the optimum attenuation level for the full state information problem and γ_N^* for the current problem $\mathcal{P}_\gamma$. Then, if C is injective, γ_N^* is continuous with respect to N at $N = 0$.*

Proof. Let $\gamma > \gamma_0^*$. There is an $\epsilon > 0$ such that Σ is defined over $[0, \epsilon]$, and sufficiently small so that $\rho(\Sigma(t)Z(t)) < \gamma^2$ over that interval. Now, looking at the equation satisfied by Σ^{-1}, it is easy to see that we may choose N small enough to ensure $\Sigma^{-1}(t) > \gamma^2 Z(t)$. Hence, for N small enough, we have $\gamma > \gamma_N^* > \gamma_0^*$. Since this was for any $\gamma > \gamma_0^*$, the proof is complete. ◇

Example. Let us return to the analysis of the running example of Chapter 4, this time with a measurement noise:

$$\dot{x} = u + w_1, \quad x_0 = 0 \text{ given},$$

$$y = x + v, \text{ and let } v = N^{\frac{1}{2}} w_2,$$

$$J = \int_0^{t_f} (x^2 + u^2) dt.$$

Recall that if we set $m = \sqrt{1-\gamma^2}/\gamma$ (hence assuming $\gamma < 1$), we have $Z(t) = (1/m)\tan[m(t_f - t)]$, and that the conjugate-point condition for Z is

$$\gamma^2 > \frac{4t_f^2}{\pi^2 + 4t_f^2}, \text{ or equivalently } t_f < \frac{\gamma}{\sqrt{1-\gamma^2}} \frac{\pi}{2}.$$

The RDE (5.28) now reads

$$\dot{\Sigma} = \left(\frac{1}{\gamma^2} - \frac{1}{N}\right)\Sigma^2 + 1, \quad \Sigma(0) = 0.$$

If the noise intensity N is very small, the coefficient $1/\gamma^2 - 1/N$ will be negative. Let $a^2 = 1/N - 1/\gamma^2$. We then have

$$\Sigma(t) = \frac{1}{a}\frac{e^{2at} - 1}{e^{2at} + 1} = \frac{1}{a}\frac{1 - e^{-2at}}{1 + e^{-2at}},$$

so that, for $N \to 0$, we have $a \to \infty$, $\Sigma(t) \to 0$ uniformly over $[0, t_f]$. Therefore, for N small enough, the conjugate-point condition on Σ will not be active, and for any fixed $\gamma > 4t_f^2/(\pi^2 + 4t_f^2)$, there is an N small enough so that condition (5.35) will be met, and hence there will be an admissible controller with attenuation level γ.

If we now fix N to be equal to 1, to make the analysis simpler, and again let $\gamma^2 < 1$, we obtain

$$\Sigma(t) = \frac{1}{m}\tan(mt).$$

Therefore, the conjugate-point conditions for Σ and Z will be the same. However, the spectral radius condition (5.35) now gives

$$\frac{1}{m^2}\tan(mt)\tan[m(t_f - t)] \le \gamma^2.$$

It is a simpler matter to check that the product is maximum for $t = t_f/2$, yielding
$$\tan^2\left(m\frac{t_f}{2}\right) \leq 1 - \gamma^2$$
or
$$t_f < \frac{2\gamma}{\sqrt{1-\gamma^2}} \tan^{-1}\left(\sqrt{1-\gamma^2}\right).$$

Now, $\sqrt{1-\gamma^2}$ is smaller than one, thus $\tan^{-1}\left(\sqrt{1-\gamma^2}\right)$ is smaller than $\pi/4$, which shows that this bound is more restrictive than the conjugate-point condition, and will actually determine the limiting γ^*. As an example, for $t_f = \pi/2$, the conjugate-point condition requires $\gamma > 1/\sqrt{2} \simeq 0.7071$, while the new condition is approximately $\gamma > 0.8524$.
◇

We end this section with a proposition showing that the above situation prevails for a large class of problems.

Proposition 5.3. *When both $Z(t)$ and $\Sigma(t)$ are positive definite (this would be true if, for example, both Q_0 and Q_f are positive definite), the spectral radius condition (5.35) is the binding constraint that determines γ^*.*

Proof. As γ is decreased to approach the conjugate-point threshold of either Z or Σ, for some t^* the corresponding matrix goes to infinity, while the other one is still bounded away from zero, so that $\rho(\Sigma(t^*)Z(t^*)) \to \infty$. Therefore condition (5.35) is violated first. ◇

5.3 Sampled-Data Measurements

We now replace the continuous measurement (5.2) by a sampled-data measurement. As in Section 4.3.2, let $\{t_k\}_{k\geq 1}$ be an increasing sequence of measurement time instants:
$$0 \leq t_1 < t_2 < \cdots < t_K < t_f.$$

Imperfect State Measurements: Continuous Time

The measurement at time t_k is of the form

$$y_k = C_k x(t_k) + E_k v_k \qquad (5.41)$$

and we take as the norm of the overall disturbance $\omega = (x_0, w, \{v_k\})$:

$$\|\omega\|^2 = |x_0|_{Q_0}^2 + \|w\|^2 + \|v\|^2$$

where

$$\|v\|^2 = \sum_{k=1}^{K} |v_k|^2.$$

We therefore have to determine whether, under the given measurement scheme, the upper value of the game with cost function

$$J_\gamma = |x(t_f)|_{Q_f}^2 + \|x\|_Q^2 + \|u\|^2 - \gamma^2 \left[|x_0|_{Q_0}^2 + \|w\|^2 + \|v\|^2\right] \qquad (5.42)$$

is bounded, and if in the affirmative, to obtain a corresponding min-sup controller

$$u(t) = \mu(t; y_1, y_2, \ldots, y_k), \quad \text{where } t_k < t \leq t_{k+1}.$$

The solution to this problem can be obtained by a simple extension of the method used in the previous section for the continuous measurement case. Toward this end, we first extend the certainty-equivalence principle. For simplicity in exposition, we will use the following compact notation which, with the exception of the last one, was introduced earlier:

$$A(t)x + B(t)u + D(t)w =: f(t; x, u, w)$$

$$C_k x + E_k v_k =: h_k(x, v_k)$$

$$|x|_{Q_f}^2 =: q(x)$$

$$-\gamma^2 |x|_{Q_0}^2 =: N(x)$$

$$|x|_Q^2 + |u|^2 - \gamma^2 |w|^2 =: g(t; x, u, w)$$

$$-\gamma^2 |v|^2 =: K(v).$$

In the sequel, we will also use the fact that $K(0) = 0$, and $K(v) < 0$, $\forall v \neq 0$.

Notice that the full state information game has, for its solution, $\hat{v} = 0$, since in that case the output plays no role, and is otherwise unchanged. Thus *Isaacs'* equation (2.16) still holds.

The auxiliary problem is as in Section 5.2, except that we must add to G^τ the terms in v:

$$G^\tau(x_0, u^\tau, \omega^\tau, v^\tau) = V(\tau; x(\tau)) + \int_0^\tau g(t; x, u, w) dt$$
$$+ \sum_{k=1}^{i(\tau)} K(v_k) + N(x_0)$$

where $i(\tau) < \tau \leq i(\tau) + 1$. Also, of course, ω now contains v, and the constraint in Ω_τ bears upon x_0, w and v. We still assume that E_k is surjective for all k.

With these changes, Theorems 5.1 and 5.2 remain intact. The proof of Theorem 5.1 is unchanged for any time instant $t \neq t_k$. At a measurement time t_k, either the new $\hat{v}_k$ is zero, and then the same argument applies (since $K(0) = 0$), or $\hat{v}_k \neq 0$, in which case W^{t_k} has a jump decrease. In all cases, $W^\tau(u, \{y_k\})$ is, for fixed $(u, \{y_k\})$, a decreasing function of τ, and the theorem holds. The proof of Theorem 5.2 equally applies here.

We solve the general auxiliary problem in a fashion similar to that of Section 5.2, except that here we must add the terms in v so that (5.23) is replaced by

$$y^\tau = \mathcal{A}^\tau u^\tau + \mathcal{B}^\tau w^\tau + \mathcal{E}^\tau v^\tau + \eta^\tau x_0 \qquad (5.43a)$$

$$z^\tau = \mathcal{C}^\tau u^\tau + \mathcal{D}^\tau w^\tau + \zeta^\tau x_0 \qquad (5.43b)$$

$$x(\tau) = \varphi^\tau u^\tau + \psi^\tau w^\tau + \Phi^\tau x_0 \qquad (5.43c)$$

where v^τ and y^τ are the sequences $\{v_k\}$ and $\{y_k\}$, truncated at $k = i$, $t_i < \tau \leq t_{i+1}$. We now need to obtain a differential equation representation of the dual operator. In this connection, Lemma 5.1 is replaced by the following:

Lemma 5.2. *The operator dual to (5.43):*

$$\alpha = \mathcal{A}^*p + \mathcal{C}^*q + \varphi^*\lambda_1$$
$$\beta = \mathcal{B}^*p + \mathcal{D}^*q + \psi^*\lambda_1$$
$$\pi = \mathcal{E}^*p$$
$$\lambda_0 = \eta^*p + \zeta^*q + \Phi^*\lambda_1$$

admits the following internal representation, in terms of a differential equation with jumps

$$-\dot{\lambda}(t) = A'(t)\lambda(t) + H'(t)q(t), \quad \lambda(t_k^-) = \lambda(t_k^+) + C_k'p_k, \quad \lambda(\tau) = \lambda_1,$$
$$\alpha(t) = B'(t)\lambda(t) + G'(t)q(t)$$
$$\beta(t) = D'(t)\lambda(t)$$
$$\pi_k = E_k p_k$$

Proof. The proof is similar to that of Lemma 5.1, except that the jumps must be taken into account, so that we have

$$\lambda'(\tau)x(\tau) - \lambda'(0)x(0) = \int_0^\tau (\dot{\lambda}'x + \lambda'\dot{x})dt + \sum_{k=1}^i (\lambda'(t_k^+) - \lambda'(t_k^-))x(t_k).$$

The criterion G^τ has the same form as in the previous case, and we just have to add the term $\mathcal{E}v$ to the constraint. Following a similar derivation, and using Lemma 5.2 to express the dual operators, finally leads to the following counterpart of equation (5.26):

$$\dot{\hat{x}}^\tau = A\hat{x}^\tau + \gamma^{-2}DD'\lambda + B\hat{u}^\tau, \quad \hat{x}^\tau(0) = \gamma^{-2}Q_0^{-1}\lambda(0), \quad (5.44a)$$
$$-\dot{\lambda} = Q\hat{x}^\tau + A'\lambda, \quad \lambda(\tau) = Z(\tau)\hat{x}^\tau(\tau), \quad (5.44b)$$
$$\lambda(t_k^-) = \lambda(t_k^+) + \gamma^2 C_k' N_k^{-1}(y_k - C_k \hat{x}^\tau(t_k)), \quad (5.44c)$$

and the expression for $\hat{u}^\tau$ is still given by (5.27). Recursive formulas for $\hat{x}^\tau(\tau)$ will be given in terms of the solution $\Sigma(t)$ of the Riccati equation with jumps:

$$\dot{\Sigma} = A\Sigma + \Sigma A' + \gamma^{-2}\Sigma Q\Sigma + DD', \quad \Sigma(0) = Q_0^{-1} \quad (5.45a)$$

$$\Sigma(t_k^+) = \Sigma(t_k^-)[I + C_k' N_k^{-1} C_k \Sigma(t_k^-)]^{-1}, \qquad (5.45b)$$

and we may notice that the jump equation (5.45b) can also be written as

$$\Sigma^{-1}(t_k^+) = \Sigma^{-1}(t_k^-) + C_k' N_k^{-1} C_k \qquad (5.45c)$$

which makes it more transparent that the positivity of $\Sigma(t)$ is preserved.

Introduce, as previously,

$$\check{x}(t) := \widehat{x}^\tau(t) - \gamma^{-2}\Sigma(t)\lambda(t), \qquad (5.46)$$

which satisfies the following differential equation with jumps, independently of τ:

$$\dot{\check{x}} = (A + \gamma^{-2}\Sigma Q)\check{x} + B\widehat{u}, \quad \check{x}(0) = 0, \qquad (5.47a)$$

$$\check{x}(t_k^+) = \check{x}(t_k^-) + \Sigma(t_k^+) C_k' N_k^{-1}(y_k - C_k \check{x}(t_k^-)), \qquad (5.47b)$$

where, due to (5.27), (5.46) and the final condition on (5.44b),

$$\widehat{u} = -B'Z(I - \gamma^{-2}\Sigma Z)^{-1}\check{x}. \qquad (5.48)$$

Two remarks should be made at this point.

Remark 5.5. As previously, if $Q = 0$ we obtain the classical sampled data Kalman filter for $\check{x}$ (see, for example, [1]). ◇

Remark 5.6. As in the Kalman filter, the equations of the sampled data measurement problem can be formally derived from those for the continuous measurement by setting in the latter

$$N^{-1}(t) = \sum_{k=1}^{K} N_k \delta(t - t_k)$$

where $\delta(\cdot)$ is the Dirac delta function. ◇

We are now in a position to state the main theorem of this section.

Theorem 5.4. Consider the disturbance attenuation problem of this section with imperfect sampled-data state measurements, and denote its optimum attenuation level by γ^*. For a given $\gamma > 0$, if the Riccati equations (5.6) and (5.45) have solutions over $[0, t_f]$, and if condition (5.35) holds (with Σ being the solution of (5.45)), then necessarily $\gamma \geq \gamma^*$. For each such γ, there exists an optimal controller, given by equations (5.45), (5.47) and (5.48). Conversely, if either of the two Riccati equations has a conjugate point or if (5.35) fails to hold, then $\gamma \leq \gamma^*$, i.e., for any smaller γ the supremum of J_γ is infinite for every admissible controller.

Proof. The proof follows the line of reasoning used in the proof of Theorem 5.3. Toward this end first assume that Z and Σ exist over $[0, t_f]$, with (5.35) satisfied. Define $W(t_f; a)$ as in (5.36), and check that now (5.37) is replaced by

$$\forall t \neq t_k, \quad k = 1, \ldots, K,$$
$$\max_w \left[\frac{\partial W}{\partial t} + \frac{\partial W}{\partial x} f(t; x, 0, w) + g(t; x, 0, w) \right] = 0,$$

and

$$\forall k = 1, \ldots, K, \quad W(t_k^-; x(t_k)) = \max_{v \mid C_k x_k + E_k v = 0} [W(t_k^+; x(t_k)) + K(v)].$$

Therefore, integrating for an arbitrary $w(\cdot)$ and $\{v_k\}$, we arrive at the same equation (5.38) as previously, leading to the same conclusion that then we do have a solution to the min-max control problem.

Proof of the necessity part of the theorem follows the same path as in Theorem 5.3, except that now the Hamiltonian system (Caratheodory's canonical equations) has to include the jumps at sampling times:

$$\dot{X} = AX + DD'Y, \quad X(0) = Q_0^{-1},$$
$$-\dot{Y} = \gamma^2 QX + A'Y, \quad Y(0) = I,$$
$$Y(t_k^+) = Y(t_k^-) + C_k' N_k^{-1} C_k X(t_k),$$

and equation (5.38) is augmented with the term $-\gamma^2 \sum_{k=1}^{i} |v_k|^2$. ◊

As a final remark, we note that for the case when x_0 is given to be *zero*, essentially the same results hold except that now $\Sigma(0) = 0$. Verification of this fact is as in Remark 5.4.

5.4 The Infinite-Horizon Case

We now turn to the stationary infinite-horizon problem, with the matrices A, B, C, D, E, and Q time invariant, (A, H) observable, (A, D) controllable, and

$$J^\infty(u,w) = \int_{-\infty}^{\infty} (|x|_Q^2 + |u|^2) dt$$

requiring that

$$x(t) \to 0 \text{ as } t \to -\infty \text{ and } t \to +\infty.$$

The theory for the full state information game, discussed in Section 4.4, remains intact, as the game over $[-T_1, T_2]$ with $x(-T_1) = \xi_1$ is the same as that over $[0, T_1 + T_2]$ with $x(0) = \xi_1$. Therefore, for any real τ, the full state information game over $[\tau, +\infty]$ has a saddle point if, and only if, the algebraic Riccati equation

$$ZA + A'Z - Z(BB' - \gamma^{-2}DD')Z + Q = 0 \tag{5.49}$$

has a (minimal) positive definite solution $\bar{Z}^+$, which is the limit integrating backward of the solution of the Riccati differential equation (5.6).[7] Otherwise the supremum is always infinite. See Section 4.5 (particularly Theorems 4.8 and 4.11) for a detailed account.

The certainty equivalence principle applies here as well. We need only consider the auxiliary problem for $\tau = 0$; Q^0, with the previous index

$$G^0(u,w) = |x(0)|_{Z^+}^2 + \int_{-\infty}^{0} (|x|_Q^2 + |u|^2 - \gamma^2|w|^2) dt$$

with $w \in \Omega_0(u, y)$, and $x(t) \to 0$ as $t \to -\infty$. (Notice that this last condition plays the role of an initial condition, uniquely specifying the trajectory associated with a given w, provided that A is stable, or stabilizable

---

[7]Henceforth, we will drop the "bar" on Z^+, for convenience in notation.

by μ). We will now bring this control problem into the form of the classical, unconstrained infinite-horizon optimization problem. Toward this end, we first introduce the variable $\theta := -t$, so that the problem is defined for $\theta \in [0, +\infty)$. In order to deal with the constraint, we make a change of coordinates in the space of the w's, letting $\bar{w}$ be the coordinates of w in Ker E, and v in (Ker $E)^\perp$. Thus there exists an orthogonal matrix P such that (recall that E is surjective)

$$w = P\begin{pmatrix} \bar{w} \\ v \end{pmatrix}, \quad EP = [0 \ E_2], \quad E_2 \text{ invertible.}$$

Let $DP =: [D_1 \ D_2]$. Since $PP' = I$, we have

$$EE' = EPP'E' = E_2 E_2' = N,$$

and hence

$$N^{-1} = (E_2^{-1})' E_2^{-1}.$$

Likewise

$$D_2 E_2^{-1} = DE'N^{-1}, \quad D_1 D_1' = DD' - DE'N^{-1}ED',$$

so that, under hypothesis (5.8b),

$$D_2 E_2^{-1} = 0, \quad D_1 D_1' = DD'.$$

The constraint now reads

$$y = Cx + E_2 v \quad \Rightarrow \quad v = E_2^{-1}(y - Cx)$$

so that, by substituting this in the dynamics and the performance index, we obtain the unconstrained problem:

$$\frac{dx}{d\theta} = -Ax - Bu - D_1 \bar{w}, \quad x(\theta) \to 0 \text{ as } \theta \to \infty$$

$$G^0 = \int_0^\infty (|x|^2_{Q-\gamma^2 C'N^{-1}C} + \gamma^2 y' N^{-1} C x$$

$$+ \gamma^2 x' C' N^{-1} y + |u|^2 - \gamma^2 |\bar{w}|^2) d\theta + |x(0)|^2_{Z^+}.$$

By the classical theory (see [88], [87]), if u and y are square integrable, G^0 has a finite supremum (with respect to $\bar{w}$) if, and only if, there exists a positive definite solution K^+ to the algebraic Riccati equation

$$KA + A'K + KDD'K + \gamma^{-2}Q - C'N^{-1}C = 0. \qquad (5.50)$$

The matrix K^+ is the maximum such solution. It is the limit, integrating backward, of the solution of the differential Riccati equation obtained by adding $dK/d\theta$ to the left-hand side of (5.50), and furthermore the matrix $-(A + DD'K^+)$ is Hurwitz.

To recover the complete formulas for this nonhomogeneous problem, we may perform again a completion of the square on the stationary Hamilton-Jacobi equation. Let

$$W(t;x) = -\gamma^2 |x - \check{x}|^2_{K^+} - \gamma^2 k(t)$$

where $\check{x}(t) \in \mathbb{R}^n$ is yet to be chosen, as well as $k(t) \in \mathbb{R}$. Recall that $\theta = -t$, so that

$$\frac{\partial W}{\partial \theta} + \frac{\partial W}{\partial x}\frac{dx}{d\theta} + g = -\frac{\partial W}{\partial t} - \frac{\partial W}{\partial x}\frac{dx}{dt} + g.$$

Straightforward manipulations show that the maximum in $\bar{w}$ in the above expression will be obtained for (we write $\hat{x}$ for the maximizing state trajectory)

$$\bar{w} = D'K^+(\hat{x} - \check{x})$$

and this maximum is zero provided that K^+ satisfies (5.50), $\check{x}$ satisfies the differential equation

$$K^+Bu + C'N^{-1}y - A'K^+\check{x} - K^+D'DK^+\check{x} - K^+\dot{\check{x}} = 0, \qquad (5.51)$$

and k is given by an integral that we need not write. (The above equation is obtained by equating the coefficient of the linear term in x to zero). Then,

$$G^0 = |\hat{x}(0)|^2_{Z^+} - \gamma^2 |\hat{x}(0) - \check{x}(0)|^2_{K^+} - \gamma^2 k(0).$$

Here, $\widehat{x}(0)$ is always part of the disturbance, since it depends on $\bar{w}$, so that G^0 has a finite supremum if, and only if,

$$Z^+ - \gamma^2 K^+ < 0, \qquad (5.52)$$

the maximum being reached for

$$Z^+\widehat{x}(0) = \gamma^2 K^+(\widehat{x}(0) - \check{x}(0)). \qquad (5.53)$$

Introduce

$$\Sigma^+ = (K^+)^{-1}.$$

This is the minimal positive definite solution of the algebraic Riccati equation

$$A\Sigma + \Sigma A' - \Sigma(C'N^{-1}C - \gamma^{-2}Q)\Sigma + DD' = 0 \qquad (5.54)$$

(as is seen using (5.50)), and can be obtained as the limit, integrating forward, of the solution of (5.28) (integrated from $\Sigma(0) = 0$ for instance).

The concavity condition (5.52) may equivalently be written as

$$Z^+ - \gamma^2(\Sigma^+)^{-1} < 0, \quad \text{or} \quad \rho(\Sigma^+Z^+) < \gamma^2. \qquad (5.55)$$

Using again (5.50) in (5.51), we arrive at

$$\dot{\check{x}} = (A + \gamma^{-2}\Sigma^+Q)\check{x} + B\widehat{u} + \Sigma^+C'N^{-1}(y - C\check{x}) \qquad (5.56)$$

which is (5.30).

The optimal (minimax) control $\widehat{u}$ will again be obtained from the certainty equivalence principle. Noting that (5.53) yields

$$\widehat{x}(0) = (I - \gamma^{-2}\Sigma^+Z^+)^{-1}\check{x}(0),$$

we obtain

$$\widehat{u} = -B'Z^+(I - \gamma^{-2}\Sigma^+Z^+)^{-1}\check{x} \qquad (5.57)$$

which is similar to (5.32). This may be substituted back in (5.56). Since we want the solution corresponding to $\check{x}(t) \to 0$ as $t \to -\infty$, $\widehat{u}$ may also be described as the output of the system with transfer function

$$T(s) = -B'Z^+(I - \gamma^{-2}\Sigma^+ Z^+)^{-1}$$
$$\cdot \left[sI - A + \Sigma^+(C'N^{-1}C - \gamma^{-2}Q) \right. \quad (5.58a)$$
$$\left. + BB'Z^+(I - \gamma^{-2}\Sigma^+ Z^+)^{-1} \right] \Sigma^+ C' N^{-1},$$

driven by y. A direct calculation allows one to see that (5.58) may also be written as:

$$T(s) = -B'Z^+ \left[sI - A + (BB' - \gamma^{-2}DD')Z^+ \right.$$
$$\left. + (I - \gamma^{-2}\Sigma^+ Z^+)^{-1}\Sigma^+ C'N^{-1}C \right]^{-1} \quad (5.58b)$$
$$\cdot \left[(I - \gamma^{-2}\Sigma^+ Z^+)^{-1}\Sigma^+ C'N^{-1} \right]$$

which corresponds to the nonstationary representation (5.33)-(5.34).

We are now in a position to state the following theorem.

Theorem 5.5. *Consider the infinite-horizon disturbance attenuation problem of this section, with optimum attenuation level γ^*. Let (A, H) be observable, $(A\ D)$ be controllable. Then, if the algebraic Riccati equations (5.49) and (5.54) both have (minimal) positive definite solutions Z^+ and Σ^+, and if they further satisfy the condition (5.55), we have $\gamma \geq \gamma^*$, and a controller ensuring such an attenuation level γ is given by (5.56)-(5.57), or equivalently by (5.33)-(5.34). The transfer function of this controller is given by any of the two expressions (5.58). If any one of the above conditions fails, then $\gamma \geq \gamma^*$.* ◊

We can now use Corollary 5.2 to derive a family of optimal (minimax) regulators. To make the statement more appealing, here we integrate over $[0, \infty)$ rather than over $(-\infty, 0]$, which would be more logical:

Corollary 5.3. *Let ψ be any causal operator from $L^2((0,\infty), \mathbb{R}^p_{N-1})$ into $L^2((0,\infty), \mathbb{R}^{m_1})$ such that, for some $\alpha \in (0,1)$,*

$$\forall y \in L^2((0,\infty), \mathbb{R}^p), \quad \int_0^\infty |\psi(y)|^2 dt \leq \alpha^2 \gamma^2 \int_0^\infty |y|^2_{N-1} dt. \tag{5.59}$$

Consider a class of controllers obtained from (5.57) by adding the term $\gamma \psi(y - \hat{y})$ as in (5.22), with ψ satisfying the additional condition that the existence and unicity of the solution to the differential equation (5.1)-(5.2) is preserved under this controller. Then, every member of this class ensures an attenuation level γ.

Proof. As in Corollary 5.2. ◇

Example (continued). We take up the illustrative example of Section 5.2, but this time with infinite horizon. It was shown in Section 4.4.3 that necessarily, $\gamma > 1$, and

$$Z^+ = \frac{\gamma}{\sqrt{\gamma^2 - 1}},$$

and for the perfect state information problem, $\gamma^* = 1$. As in Section 5.2, the equation for Σ is the same as that for Z. Hence, we have

$$\Sigma^+ = Z^+$$

and the global concavity condition yields

$$\Sigma^+ Z^+ = \frac{\gamma^2}{\gamma^2 - 1} < \gamma^2$$

hence $\gamma^2 > 2$. Therefore, for this imperfect information problem, we have $\gamma^* = \sqrt{2}$, which shows an increase in the achievable attenuation level, due to the presence of disturbance in the measurement equation.

5.5 More General Classes of Problems

5.5.1 Cross terms in the cost function

We investigate here how the results of the previous sections generalize to cases where the simplifying assumptions (5.4c) and (5.8b) do not hold. Consider a problem with a cost function of the form

$$J = |x(t_f)|^2_{Q_f} + \int_0^{t_f} (x' \ u') \begin{pmatrix} Q & P \\ P' & R \end{pmatrix} \begin{pmatrix} x \\ u \end{pmatrix} dt, \tag{5.60a}$$

where the weighting matrix of the second term is taken to be nonnegative definite, and $R > 0$. Note that in this formulation (5.4b) and (5.4c) have respectively been replaced by

$$G'(t)G(t) =: R, \tag{5.60b}$$

$$H'(t)G(t) =: P. \tag{5.60c}$$

Symmetrically, we shall have, instead of (5.8b),

$$D(t)E'(t) =: L. \tag{5.61}$$

As seen in Section 4.5.1, the cross term in the cost is most easily dealt with by making a substitution in the form[8] $u = \bar{u} - R^{-1}P'x$. It is also convenient to introduce

$$\bar{A} = A - BR^{-1}P', \quad \bar{Q} = Q - PR^{-1}P' \tag{5.62}$$

so that the RDE (5.6) becomes

$$\dot{Z} + Z\bar{A} + \bar{A}'Z - Z(BR^{-1}B' - \gamma^{-2}DD')Z + \bar{Q} = 0, \quad Z(t_f) = Q_f \tag{5.63}$$

and the minimax full state information control for a fixed value of γ is

$$u^* = \mu^*(x) = -R^{-1}(B'Z + P')x.$$

Similarly, for the estimation part it will be useful to introduce

$$\widetilde{A} = A - LN^{-1}C, \quad \widetilde{M} = DD' - LN^{-1}L'. \tag{5.64}$$

---

[8] Note that, as compared with the notation of Section 4.5.1, we have $\bar{u} = R^{-\frac{1}{2}}\tilde{u}$, $\bar{A} = \widetilde{A}$, $\bar{Q} = \widetilde{Q}$.

We may either keep the cross terms in the derivation of Theorem 5.3, as in [23], or use the transformation we used in the previous section, to obtain the generalized version of the RDE (5.28), which becomes

$$\dot{\Sigma} = \tilde{A}\Sigma + \Sigma\tilde{A}' - \Sigma(C'N^{-1}C - \gamma^{-2}Q)\Sigma + \widetilde{M}, \quad \Sigma(0) = Q_0^{-1}, \quad (5.65)$$

and of the compensator (5.30), which is now

$$\dot{\check{x}} = (\tilde{A} + \gamma^{-2}\Sigma Q)\check{x} + (\Sigma C' + L)N^{-1}(y - C\check{x}) + (B + \gamma^{-2}\Sigma P)\hat{u}. \quad (5.66a)$$

An alternative form for this equation, also valid in the former case, of course, but more appealing here, is

$$\dot{\check{x}} = A\check{x} + B\hat{u} + \gamma^{-2}\Sigma H'\check{z} + (\Sigma C' + L)N^{-1}(y - C\check{x}) \quad (5.66b)$$

where

$$\check{z} := H\check{x} + G\hat{u},$$

and

$$\hat{u} = -R^{-1}(B'Z + P')(I - \gamma^{-2}\Sigma z)^{-1}\check{x}. \quad (5.67)$$

The other form, (5.33)-(5.34), also takes on a rather simple form:

$$\dot{\hat{x}} = A\hat{x} + B\hat{u} + D\hat{w} - (I - \gamma^{-2}\Sigma Z)^{-1}(\Sigma C' + L)N^{-1}(y - \hat{y}). \quad (5.68)$$

$$\hat{u} = -R^{-1}(B'Z + P')\hat{x} \quad (5.69)$$

where we have introduced

$$\hat{w} = \gamma^{-2}D'Z\hat{x}, \quad \hat{y} = C\hat{x} + E\hat{w}.$$

Finally, in Corollary 5.2, or Corollary 5.3, we must use $\psi(y - (C + \gamma^{-2}L'Z)\hat{x})$, and take into account the weighting matrix R on u, leading to the following counterpart of condition (5.40):

$$\forall \tau \in [0, t_f], \exists \alpha \in (0,1) : \forall y(\cdot) \in L^2, \int_0^\tau |\psi(y)|_{R^{-1}}^2 dt \leq \alpha \int_0^\tau |y|_{N^{-1}}^2 dt \quad (5.70)$$

We may now summarize the above in the following Theorem.

Theorem 5.6. *Consider the disturbance attenuation problem of this section, with the system described by (5.1)-(5.2), and the performance index given by (5.60). Let its optimum attenuation level be γ^*.*

i) *Using the notation introduced by (5.8a), (5.61), (5.62) and (5.64), if the RDE's (5.63) and (5.65) have solutions Z and Σ defined over $[0, t_f]$, that meet condition (5.35), then $\gamma \geq \gamma^*$.*

ii) *A family of controllers that ensure such an attenuation level γ is given by $u = \widehat{u} + \gamma \psi(y - \widehat{y})$, where $\widehat{u}$ is given by (5.66) and (5.67), or equivalently by (5.68) and (5.69), and ψ is any causal operator satisfying condition (5.70), and preserving existence and unicity of the solution to the differential equation (5.1).*

iii) *If either (5.63) or (5.65) has a conjugate point in $[0, t_f]$, or if (5.35) fails to hold, then $\gamma \leq \gamma^*$.*

iv) *If the initial state x_0 is known to be zero, then the initial condition in (5.65) must be replaced by $\Sigma(0) = 0$, and the above holds provided that the pair (A, D) is completely reachable over $[\tau, t_f]$ for every $\tau \in [0, t_f]$.*

v) *For the infinite-horizon version, the solution is obtained by replacing the conditions on the existence of solutions to the two RDE's by conditions on the existence of their (positive definite) limits Z^+ as $t \to -\infty$ and Σ^+ as $t \to +\infty$, and replacing Z and Σ by Z^+ and Σ^+ in condition (5.35) and in (5.66b)-(5.67) or (5.68)-(5.69).* ◇

5.5.2 Delayed measurements

We now consider the class of problems where the available information at time t is only $y_{[0, t-\theta]} =: y^{t-\theta}$ where $\theta > 0$ is a time delay. As long as $t < \theta$, there is no available information, and the situation is the same as in the case investigated in Section 4.2.4. Then, we can easily extend the certainty equivalence principle to that situation, the constraint in the auxiliary problem being

$$\omega^\tau \in \Omega_{\tau-\theta}^\tau.$$

Since there is no constraint on $w(t), t \in [\tau - \theta, \tau]$, by a simple dynamic programming argument, we easily see that in that time interval, $\widehat{w}(t) = \nu^*(t; \widehat{x}(t))$, and $\widehat{u}(t) = \mu^*(t; \widehat{x}(t))$. As a consequence, we shall have

$$V(\tau, x(\tau)) + \int_{\tau-\theta}^\tau g(t; x, \widehat{u}, \widehat{w}) dt = V(\tau - \theta, x(\tau - \theta))$$

so that

$$\sup_{\omega^\tau \in \Omega_{\tau-\theta}^\tau} G^\tau(u^\tau, \omega^\tau) = \sup_{\omega^{\tau-\theta} \in \Omega_{\tau-\theta}^{\tau-\theta}} G^{\tau-\theta}(u^{\tau-\theta}, \omega^{\tau-\theta}). \qquad (5.71)$$

The trajectory of the delayed information auxiliary problem is therefore obtained by solving the standard auxiliary problem up to time $t = \tau - \theta$, and then using $u(t) = \mu^*(t, \widehat{x}(t))$, and $w(t) = \nu^*(t; \widehat{x}(t))$. Of course, this assumes that the Riccati equation (5.6) has a solution over $(0, \tau)$.

Now, the concavity conditions for the new auxiliary problem comprises two parts. Firstly, the problem from $\tau - \theta$ to τ, with $u(t)$ fixed as an open-loop policy as above, must be concave. We have seen in Section 4.2.1 (cf. Theorem 4.1), as also discussed in Section 4.2.4, that this is so if the Riccati equation

$$\dot{S} + A'S + SA + \gamma^{-2}SDD'S + Q = 0, \quad S(\tau) = Z(\tau)$$

has a solution over $[\tau - \theta, \tau]$. Secondly, we have the conditions of the full state delayed information problem, which allow us to bring concavity back to that of problem $G^{\tau-\theta}$ in (5.71), in view of Theorem 5.3. Hence, we arrive at the following result, which is essentially obtained by applying a certainty equivalence principle to the solution given in Theorem 4.7 for the full state delayed information problem.

Theorem 5.7. *Consider the problem with delayed imperfect information, with the delay characterized by the constant θ. Let its optimum attenuation level be denoted by γ^*. Given a $\gamma > 0$, if the conditions for the full state delayed information problem (cf. Theorem 4.4 or Theorem 4.7) are satisfied, and if furthermore equation (5.28) has a solution over $[0, t_f - \theta]$, satisfying (5.35) over that interval, then necessarily $\gamma \geq \gamma^*$. Moreover, an optimal controller is obtained by placing $\hat{x}^{t-\theta}(t-\theta)$ instead of $x(t-\theta)$ in the solution (4.19a)-(4.20), where $\hat{x}^{t_f}(t)$ is given by (5.33).* ◇

5.5.3 Nonlinear/nonquadratic problems

Let us take up the nonlinear/nonquadratic problem of Section 4.6, as defined by (4.54), (4.55a), but with a nonlinear imperfect measurement as in (5.12b). We may again assume a quadratic measure of the intensity of the noise, and of the amplitude of the initial state. (Although at this level of generality this might as well be replaced by any other integral measure).

For a given γ, assume that the associated soft-constrained differential game with CLPS information admits a saddle point, with a value function $V(t;x)$, and a feedback minimax strategy $\mu_\gamma(t;x)$.

We have stressed in Remark 5.2 that the certainty equivalence principle holds regardless of the linear-quadratic nature of the problem at hand, provided that the conditions for the application of Danskin's theorem (cf. Appendix B) are satisfied. We may therefore still define the auxiliary problem as in (5.13)-(5.16), and state Theorems 5.1 and 5.2.

Furthermore, notice that the function $V(t;x)$ can be computed off-line as a preliminary step, while a possible tool to solve the auxiliary problem is the forward Hamilton Jacobi equation (5.37) reformulated here as

$$\max_{w \mid h(t;x,w)=y(t)} \left[\frac{\partial W}{\partial t} + \frac{\partial W}{\partial x} f(t;x,u,w) + g(t;x,u,w) \right] = 0,$$

$$W(0;x) = N(x).$$

Imperfect State Measurements: Continuous Time

In principle, this equation can be solved recursively in real time. Assume, for instance, that the state space has been discretized with a finite mesh, and $W(t;x)$ is represented by the vector $\widehat{W}(t)$ of its values at the discretization points. Assume further that $\partial W/\partial x$ is approximated by finite differences. Then, this is a forward differential equation driven by y, of the form

$$\frac{d\widehat{W}(t)}{dt} = F(t;\widehat{W}(t),u(t),y(t)).$$

Finally, we can, in principle, solve in real time for

$$\hat{x}^t(t) = \arg\max_x [V(t;x) - W(t;x)].$$

If W is well defined and the above argmax is unique, then by Theorem 5.1, a minimax strategy is $u(t) = \mu_\gamma(t;\hat{x}^t(t))$.

5.6 Main Results of the Chapter

This chapter has presented counterparts of the results of Chapter 4 when the measurements available to the controller are disturbance corrupted. For the basic finite-horizon problem where the measurements are of the type (5.2), the characterization of a near-optimal controller is given in terms of two Riccati differential equations, one of which is the RDE encountered in Chapter 4 (see (5.6)) and the other one is a "filtering" Riccati equation which evolves in forward time (see (5.28)). The controller features a certainty equivalence property, reminiscent of the standard LQG regulator problem (see (5.27), or (5.32)), and it exists provided that, in addition to the condition of nonexistence of a conjugate point to the two RDE's, a spectral radius condition on the product of the two solutions of the RDE's (see (5.35a)) is satisfied. The precise statement for this result can be found in Theorem 5.3, whose counterpart for the infinite-horizon case (which is the original *four-block problem*) is Theorem 5.5.

In addition to the standard CLIS information structure, the chapter has also presented results on the imperfect sampled-data measurement case (see

Theorem 5.4), and delayed imperfect measurement case (see Theorem 5.7), both on a finite time horizon, with the controller (guaranteeing an attenuation level $\gamma > \gamma^*$) in each case satisfying a form of the certainty equivalence principle.

The results and the derivation presented here for the linear problem are based on the work reported in [23], [25], and this methodology has some potential applications to nonlinear/nonquadratic problems as well (as discussed in Section 5.5.3). The basic result on the infinite-horizon problem with continuous measurements (cf. Theorem 5.5; with two ARE's and a spectral radius condition) was first derived in [37], using a method quite different than the one presented here. Since then, several other derivations and extensions have appeared, such as those reported in [80], [55], [50], [51], [71], [59], [81], [82], where the last two employ a direct "completion of squares" method. Some extensions to systems with multiple controllers, using decentralized measurements, can be found in [83]. The idea of breaking the time interval into two segments and looking at two different dynamic optimization problems, one in forward and the other one in backward time (as used here) was also employed in [50]; for an extensive study of the backward/forward dynamic programming approach applied to other types of decision problems the reader is referred to [85]. The results on the sampled-data and delayed measurement cases appear here for the first time.

Chapter 6

The Discrete-Time Problem With Imperfect State Measurements

6.1 The Problem Considered

We study in this chapter the discrete-time counterpart of the problem of Chapter 5; said another way, the problem of Chapter 3, but with the measured output affected by disturbance:

$$x_{k+1} = A_k x_k + B_k u_k + D_k w_k, \qquad (6.1)$$

$$y_k = C_k x_k + E_k w_k. \qquad (6.2)$$

The cost function for the basic problem is the same as in (3.2a):

$$L(u, w) = |x_{K+1}|^2_{Q_f} + \sum_{k=1}^{K} |x_k|^2_{Q_k} + |u_k|^2 \equiv |x_{K+1}|^2_{Q_f} + \|x\|^2_Q + \|u\|^2 \quad (6.3a)$$

that we shall sometimes write as

$$J(x_1; u, w) = |x_{K+1}|^2_{Q_f} + \|z\|^2, \qquad (6.3b)$$

where z is the "controlled output":

$$z_k = H_k x_k + G_k u_k, \qquad (6.4)$$

with

$$H'_k H_k = Q_k, \qquad (6.5a)$$

$$G'_k G_k = I, \qquad (6.5b)$$

$$H'_k G_k = 0. \tag{6.5c}$$

For ease of reference, we quote the solution given in Section 3.3 for the full state-information min-max design problem with attenuation level γ. Let $\{M_k\}_{k \leq K+1}$ be the solution of the discrete-time Riccati equation (3.4), that we shall write here in one of its equivalent forms:

$$M_k = A'_k(M_{k+1}^{-1} + B_k B'_k - \gamma^{-2} D_k D'_k)^{-1} A_k + Q_k, \quad M_{K+1} = Q_f. \tag{6.6}$$

This form assumes that M_k is positive definite. As mentioned in Section 3.2.1, this is guaranteed by the following hypothesis that we shall make throughout the present chapter: $Q_f > 0$, and

$$\text{rank} \begin{bmatrix} A_k \\ H_k \end{bmatrix} = n, \tag{6.7a}$$

(i.e., $[A'_k \ H'_k]'$ is injective). Likewise, we shall assume that

$$\text{rank} [A_k \ D_k] = n \tag{6.7b}$$

(i.e., $[A_k \ D_k]$ is surjective). The hypotheses (6.7a) and (6.7b) are less restrictive than assuming complete observability in the first case, and complete reachability in the second.

Under the first hypothesis, the optimal full state information controller may be written as

$$u_k = \mu_k^*(x_k) = -B'_k(M_{k+1}^{-1} + B_k B'_k - \gamma^{-2} D_k D'_k)^{-1} A_k x_k. \tag{6.8}$$

In the present chapter, we assume that only imperfect information on the state x_k is available. More precisely, in the standard problem we shall assume that the information available to the controller at time k is the sequence $\{y_i, i = 1, \ldots, k-1\}$, that we shall write as y^{k-1}, or $y_{[1,k-1]}$. Hence, admissible controllers will be of the form

$$u_k = \mu_k(y^{k-1}) \tag{6.9}$$

and the set of all such controllers will be denoted by $\mathcal{M}$. As in the continuous-time case, we shall assume that E_k is surjective, and let

$$N_k = E_k E'_k \tag{6.10}$$

which is therefore invertible. Dually to (6.5c), we shall also assume

$$D_k E'_k = 0 \tag{6.11}$$

to simplify the derivation. After solving the problem under these restrictions, we will discuss in Section 6.3 extensions to cases where (6.5c) and/or (6.11) are not satisfied.

In the standard problem, x_1 will be part of the disturbance, and we shall use the notation w for $\{w_k\}_{k \geq 1}$, and

$$(x_1, w) =: \omega \in \Omega := \mathbb{R}^n \times \mathcal{W}.$$

This extended disturbance formulation is convenient to work with, because it allows for derivations parallel to the continuous-time case, already discussed in Chapter 5. However, as pointed out in Chapter 3 (Section 3.5.2), this case can be embedded in the fixed initial state case by including one more time step and defining

$$x_1 = w_0, \quad x_0 = 0.$$

As in the continuous-time case, we shall show during the course of the derivation how the result should be modified if instead x_1 is taken to be zero. To obtain relatively simpler results, this will require the complete reachability of (A_k, D_k) over $[1, k]$, $k = 2, \ldots$, which is equivalent to (6.7b) together with D_1 being surjective.

Introduce the extended performance index

$$J_\gamma(u, \omega) = |x_{K+1}|^2_{Q_f} + \|x\|^2_Q + \|u\|^2 - \gamma^2 \|w\|^2 - \gamma^2 |x_1|^2_{Q_0} \tag{6.12}$$

with Q_0 positive definite. The problem to be solved is then the following:

Problem $\mathcal{P}_\gamma$. Obtain necessary and sufficient conditions on γ for the upper value of the game with kernel (6.12),

$$\inf_{\mu \in \mathcal{M}} \sup_{\omega \in \Omega} J_\gamma(\mu, \omega),$$

to be finite, and for such a γ find a controller under which this upper value (which is zero) is achieved. The infimum of all such γ's will be denoted by γ^*. ◇

The solution to this problem is given in Section 6.2, by employing a forward-and-backward dynamic programming approach.[1] In Section 6.3, we solve the stationary infinite-horizon problem, which corresponds to the discrete-time four-block H^∞-optimal control problem.

In Section 6.4, we give the formulas for the case where the restriction (6.11) does not apply. We also generalize the "one-step predictor" imposed by the form (6.10) to an arbitrary "θ steps predictor", including the case $\theta = 0$, i.e., the "filtering" problem. Furthermore we include a discussion on the nonlinear problem. The chapter concludes with Section 6.5, which summarizes the main points covered.

6.2 A Certainty Equivalence Principle and Its Application to the Basic Problem $\mathcal{P}_\gamma$

To simplify the notation, we shall rewrite (6.1), (6.2) as

$$x_{k+1} = f_k(x_k, u_k, w_k) \tag{6.13a}$$

$$y_k = h_k(x_k, w_k) \tag{6.13b}$$

and (6.12) as

$$J_\gamma = M(x_{K+1}) + \sum_{k=1}^{K} g_k(x_k, u_k, w_k) + N(x_1).$$

For fixed sequences $\bar{u}^{\tau-1} := (\bar{u}_1, \bar{u}_2, \ldots, \bar{u}_{\tau-1})$ and $\bar{y}^\tau := (\bar{y}_1, \bar{y}_2, \ldots, \bar{y}_\tau)$, introduce the constraint set

$$\Omega_\tau(\bar{u}^{\tau-1}, \bar{y}^\tau) = \{\omega \in \Omega \mid y_k = \bar{y}_k, k = 1, \ldots, \tau\}. \tag{6.14}$$

---

[1] For an extensive discussion of this approach, as it arises in other contexts, see [85].

Here, y_k is the output generated by $\bar{u}$ and w, and the constraint may be checked knowing only $w^\tau := (x_1, w_1, w_2, \ldots, w_\tau)$. In fact, we shall need to consider sequences of length $\tau' \geq \tau$, and we shall write

$$\Omega_\tau^{\tau'} = \left\{ w^{\tau'} \in \Omega^{\tau'} \mid y_k = \bar{y}_k, k = 1, \ldots, \tau \right\}. \tag{6.15}$$

Let the value function of the full state-feedback two-person dynamic game defined by (6.1) and (6.13), i.e., $|x_k|^2_{M_k}$ be denoted $V_k(x_k)$. Introduce the auxiliary performance index

$$G^\tau(u^\tau, w^\tau) = V_{\tau+1}(x_{\tau+1}) + \sum_{k=1}^{\tau} g_k(x_k, u_k, w_k) + N(x_1) \tag{6.16}$$

and consider the auxiliary problem $Q^\tau(\bar{u}^{\tau-1}, \bar{y}^{\tau-1})$:

$$\max_{w^{\tau-1} \in \Omega_{\tau-1}^{\tau-1}(\bar{u}^{\tau-2}, \bar{y}^{\tau-1})} G^{\tau-1}(\bar{u}^{\tau-1}, w^{\tau-1}) \tag{6.17}$$

We may now introduce the controller which will turn out to be optimal (min-max) for $\mathcal{P}_\gamma$. Let $\hat{w}^\tau$ be the solution of the auxiliary problem Q^τ, and $\hat{x}^\tau$ be the trajectory generated by $\bar{u}^{\tau-1}$ and $\hat{w}^\tau$. The controller we will use is

$$u_\tau = \mu_\tau^*(\hat{x}_\tau^\tau) = \hat{\mu}_\tau(\bar{u}^{\tau-1}, \bar{y}^{\tau-1}). \tag{6.18}$$

This defines a complete sequence $u = \hat{\mu}(y)$, where $\hat{\mu}$ is strictly causal.

The intuition is clear. The above says that at time τ, knowing y_k up to $k = \tau - 1$, one should look for the worst possible disturbance compatible with the available information, compute the corresponding current x_τ, and "play" as if the current state were actually that most unfavorable one.

We may now state the main theorem:

Theorem 6.1. *If, for every $y \in \mathcal{Y}$, and every $\tau \in [1, K]$, the auxiliary problem $Q^\tau(\hat{\mu}(y), y)$ has a unique maximum attained for $w = \hat{w}^\tau$, generating a state trajectory $\hat{x}^\tau$, then (6.18) defines a min sup controller for $\mathcal{P}_\gamma$, and the min-max cost is*

$$\min_{\mu \in \mathcal{M}} \max_{w \in \Omega} J_\gamma = \max_{x_1 \in \mathbb{R}^n} (V_1(x_1) + N(x_1)) = V_1(\hat{x}_1^1) + N(\hat{x}_1^1).$$

Proof. We need to consider an auxiliary problem where $u^{\tau-1}$ is fixed at $\bar{u}^{\tau-1}$, and u_τ is given *a priori* as a function of x_τ, as $u_\tau = \mu_\tau^*(x_\tau)$. Let $\bar{u}^{\tau-1} \cdot \mu_\tau^*$ be that strategy.

Lemma 6.1. *We have*

$$\max_{\omega^{\tau-1} \in \Omega^{\tau-1}_{\tau-1}(\bar{u}^{\tau-2}, \bar{y}^{\tau-1})} G^{\tau-1}(\bar{u}^{\tau-1}, \omega^{\tau-1})$$

$$= \max_{\omega^\tau \in \Omega^{\tau}_{\tau-1}(\bar{u}^{\tau-2}, \bar{y}^{\tau-1})} G^\tau(\bar{u}^{\tau-1} \cdot \mu_\tau^*, \omega^\tau),$$

and the optimal $\hat{\omega}^{\tau-1}$ of both problems coincide.

Proof. (Notice that this is the property (5.71) we used in Section 5.5.2). In the problem of the right-hand side, w_τ is unconstrained, since it influences only $x_{\tau+1}$ and y_τ. Hence, by dynamic programming, w_τ maximizes $V_{\tau+1}(x_{\tau+1}) + g(x_\tau, \mu_\tau^*(x_\tau), w_\tau)$, and by *Isaacs'* equation (see (2.12)) we know that the maximizing w_τ is $\nu_\tau^*(x_\tau)$, and the resulting value of this sum is $V_\tau(x_\tau)$. ◇

Continuing with the proof of Theorem 6.1, we may now define

$$W^\tau(\bar{u}^{\tau-1}, \bar{y}^{\tau-1}) := \max_{\omega^{\tau-1} \in \Omega^{\tau-1}_{\tau-1}(\bar{u}^{\tau-2}, \bar{y}^{\tau-1})} G^{\tau-1}(\bar{u}^{\tau-1}, \omega^{\tau-1})$$

$$= \max_{\omega^\tau \in \Omega^\tau_{\tau-1}(\bar{u}^{\tau-2}, \bar{y}^{\tau-1})} G^\tau(\bar{u}^{\tau-1} \cdot \mu_\tau^*, \omega^\tau).$$

Using the representation theorem (Theorem 2.5), we also have, freezing u_τ:

$$W^\tau(\bar{u}^{\tau-1}, \bar{y}^{\tau-1}) = \max_{\omega^\tau \in \Omega^\tau_{\tau-1}(\bar{u}^{\tau-2}, \bar{y}^{\tau-1})} G^\tau(\bar{u}^{\tau-1} \cdot \mu_\tau^*(\hat{x}_\tau^\tau), \omega^\tau).$$

(On the right-hand side, $\hat{x}_\tau^\tau$ is fixed, as generated by the $\hat{\omega}^\tau$ solution of Q^τ).

Assume therefore that $\bar{u}_\tau = \hat{\mu}(\hat{x}_\tau^\tau)$. Since

$$\Omega_\tau^\tau(\bar{u}^{\tau-1}, \bar{y}^\tau) \subset \Omega^\tau_{\tau-1}(\bar{u}^{\tau-2}, \bar{y}^{\tau-1}),$$

we get, with this controller, provided that $G^\tau(\bar{u}^\tau, \omega^\tau)$ is concave,

$$W^{\tau+1}(\bar{u}^\tau, \bar{y}^\tau) = \max_{\omega^\tau \in \Omega^\tau_\tau(\bar{u}^{\tau-1}, \bar{y}^\tau)} G^\tau(\bar{u}^{\tau-1} \cdot \mu_\tau^*(\hat{x}_\tau^\tau), \omega^\tau)$$

$$\leq W^\tau(\bar{u}^{\tau-1}, \bar{y}^{\tau-1}).$$

(6.19)

Imperfect State Measurements: Discrete Time

We conclude that if $u = \hat{\mu}(y)$, W^τ decreases with τ, and that

$$\forall y, \quad W^{K+1}(\hat{\mu}(y),y) \leq W^1(\hat{\mu}_1(y_1),y_1) = \max_{x_1 \in \mathbf{R}^n} (V_1(x_1) + N(x_1)).$$

Furthermore, since $G^K(u,\omega) = J_\gamma(u,\omega)$, and for any $\omega \in \Omega$, if y is the output it generates with (6.18), then necessarily $\omega \in \Omega_K(\hat{\mu}(y),y)$, we get

$$J_\gamma(\hat{\mu}(y),\omega) \leq V_1(\hat{x}_1^1) + N(\hat{x}_1^1). \tag{6.20}$$

Now, it is possible for the maximizing player to choose the initial state $\hat{x}_1^1$ and then the w coinciding with $\nu^*(x)$, and this ensures the bound:

$$\forall u, \quad J_\gamma(u,\omega^*) \geq V_1(\hat{x}_1^1) + N(\hat{x}_1^1). \tag{6.21}$$

A comparison of (6.20b) and (6.21) establishes the result. ◇

The above derivation shows that whatever $\bar{u}^{\tau-1}$ and $\bar{y}^{\tau-1}$ are, if from time τ onwards we use (6.18), W^τ decreases. Hence we have in fact established a stronger result than that claimed in Theorem 6.1:

Corollary 6.1. *For any given τ and any past u^τ, and under the hypotheses of Theorem 6.1, using the controller $\hat{\mu}$ of (6.18) will ensure*

$$J_\gamma \leq W^\tau(u^\tau, y^\tau), \tag{6.22}$$

and this is the best (least) possible guaranteed outcome. ◇

We now have the following theorem, which is important in the context of our problem $\mathcal{P}_\gamma$.

Theorem 6.2. *If for all $(u,y) \in \mathcal{U} \times \mathcal{Y}$ there exists a k^* such that problem Q^{k*} defined by (6.17) fails to have a finite supremum, then problem $\mathcal{P}_\gamma$ has no solution, the supremum with respect to ω being infinite for any controller $\mu \in \mathcal{M}$.*

Proof. The proof is identical to that of Theorem 5.2. ◇

Remark 6.1. Theorems 6.1 and 6.2 do not depend on the linear-quadratic character of problem $\mathcal{P}_\gamma$, and Theorem 6.1, in particular, might be interesting in its own sake since it applies to problems with nonlinear state dynamics and/or nonquadratic cost functions. However, the use we shall now make of them in the sequel is completely dependent on the specific (linear-quadratic) form of $\mathcal{P}_\gamma$. ◇

To make things simpler, we reformulate problem $\mathcal{Q}^\tau$ in a more compact way. Let $u^{\tau-1} \in \ell^2([1,\tau-1], I\!R^{m_1}) \sim I\!R^{m_1(\tau-1)}$, $y^{\tau-1} \in \ell^2([1,\tau-1], I\!R^p) \sim I\!R^{p(\tau-1)}$, $z^{\tau-1} \in \ell^2([1,\tau-1], I\!R^q) \sim I\!R^{q(\tau-1)}$, and $\mathcal{A}^{\tau-1}$, $\mathcal{B}^{\tau-1}$, $\mathcal{C}^{\tau-1}$, $\mathcal{D}^{\tau-1}$, $\eta^{\tau-1}$, $\zeta^{\tau-1}$, $\varphi^{\tau-1}$, $\psi^{\tau-1}$, and $\Phi^{\tau-1}$ be linear operators, such that (6.1), (6.2), (6.4) considered over $[1, \tau-1]$ may be written as in (5.23):

$$y^{\tau-1} = \mathcal{A}^{\tau-1} u^{\tau-1} + \mathcal{B}^{\tau-1} w^{\tau-1} + \eta^{\tau-1} x_1 \qquad (6.23a)$$

$$z^{\tau-1} = \mathcal{C}^{\tau-1} u^{\tau-1} + \mathcal{D}^{\tau-1} w^{\tau-1} + \zeta^{\tau-1} x_1 \qquad (6.23b)$$

$$x_\tau = \varphi^{\tau-1} u^{\tau-1} + \psi^{\tau-1} w^{\tau-1} + \Phi^{\tau-1} x_1. \qquad (6.23c)$$

For the sequel of this derivation, we shall omit the superscript $\tau - 1$, for simplicity. We have the following fact:

Lemma 6.2. *The operator dual to (6.23):*

$$\begin{aligned}
\alpha &= \mathcal{A}^* p + \mathcal{C}^* q + \varphi^* \lambda_f, \\
\beta &= \mathcal{B}^* p + \mathcal{D}^* q + \psi^* \lambda_f, \\
\lambda_1 &= \eta^* p + \zeta^* q + \Phi^* \lambda_f,
\end{aligned}$$

admits the following internal representation, giving λ_1 and the sequences α and β:

$$\lambda_k = A'_k \lambda_{k+1} + C'_k p_k + H'_k q_k, \quad \lambda_\tau = \lambda_f$$

$$\alpha_k = B'_k \lambda_{k+1} + G'_k q_k,$$
$$\beta_k = D'_k \lambda_{k+1} + E'_k p_k.$$

Proof. Write

$$(\lambda_\tau, x_\tau) - (\lambda_1, x_1) = \sum_{k=1}^{\tau-1} (\lambda_{k+1}, x_{k+1}) - (\lambda_k, x_k),$$

and substitute for λ_k using the above equation and for x_{k+1} using (6.1). Add and subtract $(G'_k q_k, u_k)$ and $(E'_k p_k, w_k)$, to obtain finally

$$(\lambda_f, x_\tau) + \sum_{k=1}^{\tau-1}(p_k, y_k) + \sum_{k=1}^{\tau-1}(q_k, z_k) = (\lambda_1, x_1) + \sum_{k=1}^{\tau-1}(\alpha_k, u_k) + \sum_{k=1}^{\tau-1}(\beta_k, w_k),$$

which proves the claim. ◇

As in Section 5.2, problem $\mathcal{Q}^\tau$ can now be stated as

$$\max_{x_1, w} \left[|\varphi u + \psi w + \Phi x_1|^2_{M_\tau} + \|\mathcal{C}u + \mathcal{D}w + \zeta x_1\|^2 - \gamma^2(\|w\|^2 + |x_1|^2_{Q_0}) \right],$$

under the constraint

$$\mathcal{A}u + \mathcal{B}w + \eta x_1 = y.$$

The derivation may be conducted exactly as in Section 5.2, *mutatis mutandis*, leading to the following necessary conditions. (Recall that we call the pair $(\widehat{\omega}^\tau, \widehat{x}^\tau)$ the solution of $\mathcal{Q}^\tau$, and $u_\tau = \widehat{u}^\tau_\tau$ is assumed to be given by (6.9) and (6.18), but be fixed in problem $\mathcal{Q}^\tau$).

$$\widehat{x}^\tau_{k+1} = A_k \widehat{x}^\tau_k + \gamma^{-2} D_k D'_k \lambda_{k+1} + B_k \widehat{u}_k, \quad \widehat{x}^\tau_1 = \gamma^{-2} Q_0^{-1} \lambda_1 \quad (6.24a)$$

$$\lambda_k = -\gamma^2 (C'_k N_k^{-1} C_k - \gamma^{-2} Q_k) \widehat{x}^\tau_k + A'_k \lambda_{k+1} + \gamma^2 C'_k N_k^{-1} y_k, \quad \lambda_\tau = M_\tau \widehat{x}^\tau_\tau. \quad (6.24b)$$

To solve this two-point boundary value problem, we introduce, when it exists, the matrix Σ_k generated by the discrete Riccati equation

$$\Sigma_{k+1} = A_k (\Sigma_k^{-1} + C'_k N_k^{-1} C_k - \gamma^{-2} Q_k)^{-1} A'_k + D_k D'_k, \quad \Sigma_1 = Q_0^{-1} \quad (6.25a)$$

(or if $x_1 = 0$ is fixed, $\Sigma_1 = 0$ and $\Sigma_2 = D_1 D_1'$, assumed positive definite). Using a standard matrix inversion lemma, (6.25a) can be given alternative forms, such as

$$\Sigma_{k+1} = A_k \Sigma_k A_k' + D_k D_k'$$

$$- A_k \Sigma_k (C_k' \; H_k') \begin{pmatrix} N_k + C_k \Sigma_k C_k' & C_k \Sigma_k H_k' \\ H_k \Sigma_k C_k' & H_k \Sigma_k H_k' - \gamma^2 I \end{pmatrix}^{-1} \quad (6.25b)$$

$$\cdot \begin{pmatrix} C_k \\ H_k \end{pmatrix} \Sigma_k A_k'.$$

Introduce also the variable

$$\check{x}_k := \widehat{x}_k^\tau - \gamma^{-2} \Sigma_k \lambda_k. \quad (6.26)$$

A direct calculation shows that $\check{x}$ satisfies the following equation:

$$\check{x}_{k+1} = A_k \check{x}_k + B_k \widehat{u}_k + A_k \left(\Sigma_k^{-1} + C_k' N_k^{-1} C_k - \gamma^2 Q_k \right)^{-1}$$
$$\cdot \left[\gamma^{-2} Q_k \check{x}_k + C_k' N_k^{-1} (y_k - C_k \check{x}_k) \right], \quad \check{x}_1 = 0. \quad (6.27)$$

Notice that (6.25)-(6.27) do not need to bear a superscript τ, since they are independent of τ, provided we know how to obtain $\widehat{u}_\tau$ as a function of $\widehat{x}_\tau$. We use the final condition of (6.24b) in (6.26), and this yields

$$\widehat{x}_\tau^\tau = (I - \gamma^{-2} \Sigma_\tau M_\tau)^{-1} \check{x}_\tau, \quad (6.28)$$

and placing this back in (6.9), according to (6.18), it becomes

$$\widehat{u}_k = -B_k' (M_{k+1}^{-1} + B_k B_k' - \gamma^{-2} D_k D_k')^{-1} A_k (I - \gamma^{-2} \Sigma_k M_k)^{-1} \check{x}_k. \quad (6.29)$$

Notice finally that existence of an $\widehat{x}_1^1$ that maximizes $V_1(x_1) + N(x_1)$ necessitates that

$$M_1 - \gamma^{-2} Q_0 = M_1 - \gamma^{-2} \Sigma_1^{-1} < 0$$

(and then $\widehat{x}_1^1 = 0$). But also, the inverse in (6.28) will exist if (this is only sufficient)

$$\forall k, \quad M_k - \gamma^{-2} \Sigma_k^{-1} < 0, \quad (6.30a)$$

or equivalently, in terms of the spectral radius $\rho(\Sigma_k M_k)$:

$$\forall k, \quad \rho(\Sigma_k M_k) < \gamma^2. \quad (6.30b)$$

Remark 6.2. If $Q = 0$, equation (6.25) and (6.27) are those of a standard discrete-time Kalman filter; see Chapter 7. ◇

We are now in a position to state the main result of this section.

Theorem 6.3. *Let the following three conditions hold:*

(i) Equation (6.6) has a solution over $[1, K+1]$, with

$$M_{k+1}^{-1} - \gamma^{-2} D_k D_k' > 0, \quad \forall k \in [1, K];$$

(ii) Equation (6.25) has a solution over $[1, K+1]$, with

$$\Sigma_k^{-1} + C_k' N_k^{-1} C_k - \gamma^{-2} Q_k > 0, \quad \forall k \in [1, K];$$

(iii) Condition (6.30a) or equivalently (6.30b).

Under these three conditions, $\gamma \geq \gamma^*$, and an optimal (minimax) controller is given by (6.6), (6.25), (6.27), and (6.29). If one of the three conditions above is not met, then $\gamma \leq \gamma^*$, i.e., for any smaller γ, problem $\mathcal{P}_\gamma$ has an infinite supremum for any admissible controller $\mu \in \mathcal{M}$.

Proof. We have seen earlier in Section 3.2 (cf. Theorem 3.2) that the first condition of the theorem is both necessary and sufficient for the full state information problem to have a solution. It is *a fortiori* necessary here. In view of Theorems 6.1 and 6.2, we must study the role of the other two conditions in the concavity of problems $\mathcal{Q}^\tau$. These are nonhomogeneous quadratic forms to be maximized under affine constraints. Strict concavity of such a problem is equivalent to that of its homogeneous part. We may therefore restrict our attention to the case where $u^{\tau-1} = 0$ and $y^{\tau-1} = 0$. We shall use a forward dynamic programming approach, with

$$W_\tau(x) = \max_{w \in \Omega_{\tau-1}^{\tau-1}(0,0)} \left[\sum_{k=1}^{\tau-1} g_k(x_k, 0, w_k) + N(x_1) \right] \text{ given } x_\tau = x, \quad (6.31)$$

and then
$$\max_{\omega \in \Omega_{\tau-1}^{\tau-1}} G^{\tau-1}(0,\omega) = \max_{x_\tau} \left[V_\tau(x_\tau) + W_\tau(x_\tau) \right]. \qquad (6.32)$$

The function W satisfies the dynamic programming equation

$$W_1(x) = N(x), \qquad (6.33a)$$

$$W_{k+1}(x_+) = \max_{x,w} \left[W_k(x) + g_k(x,0,w) \right] \text{ subject to } \begin{cases} x_+ = f_k(x,0,w) \\ 0 = h_k(x,w) \end{cases} \qquad (6.33b)$$

Because of the linear-quadratic nature of the problem, we solve it with

$$W_k(x) = |x|^2_{K_k}. \qquad (6.34)$$

Let us investigate one step of the forward dynamic programming. To simplify the notation, we shall write x and w for x_k and ω_k, x_+ for x_{k+1}, and likewise K and K_+ for K_k and K_{k+1}.

Equations (6.1) and (6.2) are now replaced by

$$x_+ = Ax + Dw, \qquad (6.35)$$

$$0 = Cx + Ew. \qquad (6.36)$$

Since E was assumed to be surjective, by an orthogonal change of basis on w, the above can be cast into the following form, where $PP' = I$, and E_2 is invertible:

$$DP = (D_1 \; D_2) \quad EP = (0 \; E_2) \quad w = P \begin{pmatrix} w_1 \\ v \end{pmatrix}. \qquad (6.37)$$

Moreover, hypothesis (6.11) yields

$$DE' = D'_2 E_2 = 0, \quad \text{and hence} \quad D_2 = 0,$$

so that (6.35)-(6.36) now read

$$x_+ = Ax + D_1 w_1 \qquad (6.38)$$

$$0 = Cx + E_2 v. \qquad (6.39)$$

In the sequel, we shall let D stand for D_1, w for w_1, and E for E_2. Notice however that $D_1 D_1' = DD'$, and $E_2 E_2' = EE' = N$. The constraint (6.39) also yields $v = -E_2^{-1} Cx$, or $|v|^2 = |x|^2_{C'N^{-1}C}$. Thus, (6.33b) now reads

$$|x_+|^2_{K_+} = \max_{x,w} \left[|x|^2_K + |x|^2_Q - \gamma^2 |w|^2 - \gamma^2 |x|^2_{C'N^{-1}C} \right]$$
(6.40)
$$\text{subject to } Ax + Dw = x_+.$$

We had assumed that $[A\ D]$ is surjective. Therefore, this problem either has a solution for every x_+, or for some x_+ it has an infinite supremum. In that second case, the problem Q^τ is unbounded. Hence a necessary condition would be that the first case prevails, leading to the conclusion that extremal trajectories may cover all of $\mathbb{R}^n$ at any time step. But now, (6.32) reads

$$\max_{x_\tau} \left[|x_\tau|^2_{M_\tau} + |x_\tau|^2_{K_\tau} \right].$$

This problem is strictly concave if, and only if,

$$M_\tau + K_\tau < 0. \qquad (6.41)$$

(And we shall see that K_τ is strictly increasing with γ, so that if $M_\tau + K_\tau$ is singular for some $\gamma > 0$, for any smaller value of γ the problem has an infinite supremum). Now, recall that under our standing hypothesis (6.7), $M_\tau > 0$. Thus, a necessary condition is

$$\forall \tau \in [2, K+1], \quad K_\tau < 0. \qquad (6.42)$$

Furthermore, this condition is also satisfied by $K_1 = -\gamma^2 Q_0$ in the standard case where x_1 is unknown.

Lemma 6.3. *Let $P := K + Q - \gamma^2 C'N^{-1}C$. A necessary condition for K_+ to be negative definite is that $P \leq 0$ and $\operatorname{Ker} P \subset \operatorname{Ker} A$*

Proof. Suppose that P is not nonpositive definite. Let x be such that $|x|^2_P > 0$.[2] If $Ax = 0$, problem (6.40) is not concave, and has an infinite

---
[2] Here, perhaps by a slight abuse of notation, we use $|x|^2_P$ to mean $x'Px$, even though it is not a norm (because of the choice of P).

supremum (since we may add ax, $a > 0$, to any (x, w) meeting the constraint). If $Ax \neq 0$, choose $x_+ = Ax$. Then $(x, 0)$ meets the constraint; thus $|x_+|_{K_+}^2 \geq |x|_P^2 > 0$. Likewise, if $|x|_P^2 = 0$, but $Ax \neq 0$, pick $x_+ = Ax$, and get $|x_+|_{K_+}^2 \geq |x|_P^2 = 0$. This completes the proof of the lemma. ◇

Thus, an orthogonal change of basis on x necessarily exists, such that

$$A = \begin{bmatrix} \widehat{A} & 0 \end{bmatrix} \quad P = \begin{bmatrix} \widehat{P} & 0 \\ 0 & 0 \end{bmatrix} \quad x = \begin{bmatrix} \widehat{x} \\ \widetilde{x} \end{bmatrix}$$

with $\widehat{P} < 0$. The maximization problem (6.40) now reads

$$\max_{\widehat{x}, w} [|\widehat{x}|_{\widehat{P}}^2 - \gamma^2 |w|^2] \quad \text{subject to} \quad x_+ = \widehat{A}x + Dw,$$

where $[\widehat{A} \ D]$ is still surjective. Introduce a Lagrange multiplier $2p \in \mathbb{R}^n$, and notice that $DD' - \gamma^2 \widehat{A}\widehat{P}^{-1}\widehat{A}' > 0$, so that the stationarity conditions

$$\widehat{P}x + \widehat{A}'p = 0, \quad -\gamma^2 w + D'p = 0, \quad \widehat{A}x + Dw = x_+$$

yield

$$\begin{aligned} x &= -\gamma^2 \widehat{P}^{-1}\widehat{A}(DD' - \gamma^2 \widehat{A}\widehat{P}^{-1}\widehat{A}')^{-1} x_+, \\ w &= D'(DD' - \gamma^2 \widehat{A}\widehat{P}^{-1}\widehat{A}')^{-1} x_+, \end{aligned}$$

and thus

$$K_+ = (\widehat{A}\widehat{P}^{-1}\widehat{A} - \gamma^{-2} DD')^{-1}. \tag{6.43}$$

We now claim the following further fact:

Lemma 6.4. *If for some k, P_k introduced earlier in Lemma 6.3 is not positive definite, then for the corresponding γ we have the inequality $\gamma \leq \gamma^*$.*

Proof. We shall prove that if P is only nonnegative definite, for any smaller γ it fails to be so, so that, according to the above, problem Q^τ fails to be concave for some τ. From (6.40),

$$|x_2|^2_{K_2} = \max_{x_1,w}\left[-\gamma^2|x_1|^2_{Q_0} + |x_1|^2_{Q_1} - \gamma^2|w_1|^2 - \gamma^2|x|^2_{C_1'N_1C_1}\right],$$

$$Ax_1 + Dw_1 = x_2.$$

Hence, here, $P_1 = -\gamma^2 Q_0 + Q_1 - \gamma^2 C_1' N_1^{-1} C_1$. Since Q_0 is positive definite, P_1 being singular for some γ implies that for any smaller value of γ it will fail to be nonpositive definite. Suppose that P_1 is negative definite. It is strictly increasing (in the sense of positive definiteness) with γ. But then according to (6.43), because $[\widehat{A}\ D]$ is surjective, K_2 is strictly increasing with γ, and recursively, so are the K_k's, as long as P is negative definite. Thus let $k = k^*$ be the first time step where P is singular. The above reasoning shows that, for any smaller γ, P will no longer be nonpositive definite. This completes the proof of the lemma. ◇

We now continue with the proof of Theorem 6.3. As a consequence of Lemma 6.4, (6.41) may be rewritten as

$$K_+ = (AP^{-1}A' - \gamma^{-2}DD')^{-1},$$

or, if we let $\Sigma_k := -\gamma^2 K_k^{-1}$, we arrive precisely at equation (6.25a), while (6.41) is then equivalent to (6.30). Moreover, the condition $P < 0$ is exactly the positivity condition stated in the theorem, so that the necessity of the three conditions to ensure $\gamma \geq \gamma^*$ is now proved. But according to (6.40), $P < 0$ is clearly sufficient to ensure concavity in the dynamic programming recursion, and therefore together with (6.30) ensures concavity of all the auxiliary problems. Thus, in view of Theorems 6.1 and 6.2, the proof of the main theorem has been completed. ◇

Remark 6.3. In the case where x_1 is given to be zero, and D_1 is surjective, only the first step differs:

$$|x_2|^2_{K_2} = \max_{w|D_1w=x_2}(-\gamma^2|w|^2) = -|x_2|^2_{\gamma^2(D_1D_1')^{-1}}$$

i.e., $K_2 = \gamma^2(D_1 D_1')^{-1}$, or $\Sigma_2 = D_1 D_1'$. Hence, we may still use equation (6.25) initialized at $\Sigma_1 = 0$, with $Q_1 = 0$. ◇

Proposition 6.1. *Let γ_0^* be the optimum attenuation level for the full state feedback problem (as given in Theorem 3.5), and γ_N^* be the one for the current (imperfect state measurement) problem with $x_1 = 0$ and $N = \{N_k\}$ given by (6.10). If the matrices C_k are injective, γ_N^* is continuous with respect to N at $N = 0$.*

Proof. According to (6.25a), if C_k is surjective, Σ_{k+1} converges to $D_k D_k'$ as $N_k \to 0$. Thus for N_k small enough, (6.25a) has a solution, and the positivity condition is met. Furthermore, let $\gamma > \gamma_0^*$. By hypothesis, the positivity condition (3.7) is satisfied, i.e.,

$$\gamma^2 I - D_k' M_{k+1} D_k > 0$$

or equivalently

$$D_k D_k' < \gamma^2 M_{k+1}^{-1}$$

so that, as Σ_{k+1} approaches $D_k D_k'$, (6.30a), at time $k+1$, will eventually be satisfied also. Therefore, for N small enough, $\gamma > \gamma_N^*$. But since γ_N^* can never be smaller that γ_0^*, we conclude that $\gamma > \gamma_N^* \geq \gamma_0^*$. Since this was true for any $\gamma > \gamma_0^*$, we conclude that $\gamma_N^* \to \gamma_0^*$ as $N \to 0$. ◇

Example. We now revisit the illustrative example of Chapter 3, by adding an observation noise:

$$x_{k+1} = x_k + u_k + w_k, \quad x_1 = 0$$
$$y_k = x_k + v_k \quad \text{where } v_k = N^{\frac{1}{2}} \widetilde{w}_k$$

Note that the disturbance is now the two-dimensional vector $(w_k \ \widetilde{w}_k)'$. The performance index is again taken to be

$$J = \sum_{k=1}^{3} x_{k+1}^2 + u_k^2 .$$

Applying the theory developed above, it is a simple matter to check that, as $N \to 0$, $\Sigma_k \to 1$ for $k = 2, 3, 4$. Since $D_k = 1$, the nonsingularity condition for the first Riccati equation is $M_k < \gamma^2$, and therefore, for N small enough, the spectral radius condition $\rho(\Sigma_k M_k) < \gamma^2$ will be satisfied.

Let us now be more explicit for the case $N = 1$. Equation (6.6) yields

$$M_4 = 1, \quad M_3 = \frac{3\gamma^2 - 1}{2\gamma^2 - 1}, \quad M_2 = \frac{3\gamma^4 - \gamma^2}{5\gamma^4 - 5\gamma^2 + 1}.$$

Since we have $\Sigma_1 = 0$, there is no condition to check on M_1. It has been seen in Section 3.2.4 that the most stringent of the positivity conditions is here $M_2 < \gamma^2$, which yields $\gamma^2 > 1.2899$ (approximately). We also find that

$$\Sigma_2 = M_4, \quad \Sigma_3 = M_3, \quad \Sigma_4 = M_2.$$

The second positivity condition of Theorem 6.3 turns out to be less stringent than its first condition (given above), and hence it does not impose any additional restriction on γ. The third (spectral radius) condition, on the other hand, yields

$$\begin{aligned}
\Sigma_2 M_2 < \gamma^2 &\iff M_2 < \gamma^2; \\
\Sigma_4 M_4 < \gamma^2 &\iff M_4 < \gamma^2; \\
\Sigma_3 M_3 < \gamma^2 &\iff M_3^2 < \gamma^2.
\end{aligned}$$

The first two are already satisfied whenever $\gamma^2 > 1.2899$. The third one brings in a more stringent condition, namely

$$\gamma > 1 + \frac{\sqrt{5}}{2} \simeq 1.6180, \text{ or } \gamma^2 > 2.6180,$$

which is the optimum attenuation level for this example. Note that, as expected, here the optimum attenuation level is higher than of the "noise-free measurement" case. ◊

6.3 The Infinite-Horizon Case

We now turn to the time-invariant infinite-horizon problem, with the pair (A, H) being observable (A, D) being controllable, and the extended performance index replaced by

$$J_\gamma^\infty(u,\omega) = \sum_{k=-\infty}^{+\infty} (|x_k|_Q^2 + |u_k|^2 - \gamma^2|w_k|^2)$$

We further require that $x_k \to 0$ as $k \to -\infty$ and as $k \to +\infty$.

We first recall from Section 3.4 (Theorem 3.7) that, for any finite integer τ, and for a fixed $\gamma > 0$, the perfect state-measurement game defined over $[\tau, +\infty)$ has a finite upper value if, and only if, the algebraic Riccati equation

$$M = A'(M^{-1} + BB' - \gamma^{-2}DD')^{-1}A + Q \tag{6.44}$$

has a (minimal) positive definite solution M^+, which is the (monotonic) limit, as $k \to -\infty$, of the solution of the Riccati difference equation (6.6) initialized at $M_0 = 0$.

The certainty equivalence principle holds intact here, using the stationary (version of) *Isaacs'* equation (2.12). We need only consider the auxiliary problem for $\tau = 0$:

$$G^{(-1)}(u,w) = |x_0|_{M^+}^2 + \sum_{k=-\infty}^{-1} |x_k|_Q^2 + |u_k|^2 - \gamma^2|w_k|^2. \tag{6.45}$$

Now, making the same change of the basis that led to (6.38)-(6.39), letting $\tilde{x}_i := x_{-i}$, and likewise for w and y, and suppressing the "tilde"s in the equations to follow, we are led to an infinite-time control problem for an implicit system, as given below:

$$Ax_{i+1} = x_i - Dw_{i+1}, \quad x_i \to 0 \text{ as } i \to \infty$$

$$G = |x_0|_{M^+}^2 + \sum_{k=1}^{\infty} (|x_k|_{Q-\gamma^2 C'N^{-1}C}^2 + \gamma^2 y_k' N^{-1} C x_k$$

$$+ \gamma^2 x_k' C' N^{-1} y_k + |u_k|^2 - \gamma^2|w_k|^2).$$

The forward dynamic programming technique of the previous section applies. Thus, we again know that the homogeneous problem has a solution for any x_0 (i.e., it is strictly concave) if, and only if, the associated algebraic

Riccati equation

$$\Sigma = A(\Sigma^{-1} + C'N^{-1}C - \gamma^{-2}Q)^{-1}A' + DD'_1, \qquad (6.46)$$

has a (minimal) positive definite solution Σ^+, or equivalently the algebraic Riccati equation

$$K^{-1} = A(K + Q - \gamma^2 C'N^{-1}C)^{-1}A' - \gamma^{-2}DD' \qquad (6.47)$$

admits a maximal negative definite solution K^-. Furthermore, the closed-loop system matrix is Hurwitz. Of course, x_0 is now a function of w, so that $G^{(-1)}$ will be concave only if, in addition, the matrix $M^+ + K^-$ is negative definite, or equivalently

$$M^+ - \gamma^{-2}(\Sigma^+)^{-1} < 0 \quad \text{or} \quad \rho(\Sigma^+ M^+) < \gamma^2. \qquad (6.48)$$

We may notice that the analysis of the asymptotic behavior of the Riccati equation (6.25a) has already been made in Section 3.4, since up to an obvious change of notation, (6.25a) is the same as (6.6). The positivity condition we impose, however, is weaker here, so that to rely on the previous analysis, we have to check that the arguments carry over. But (6.40) still shows that K_+ is an increasing function of the number of stages in the game, and the positivity condition (3.7), invoked in Lemma 3.3, was in fact used only to show that the matrix $M_{k+1}^{-1} + B_k B'_k - \gamma^{-2} D_k D'_k$ is positive definite, and this corresponds to the positivity condition here. So we may still rely here on the derivation of Chapter 3 (Section 3.4).

To recover the nonhomogeneous terms, let

$$W_k(x) = |x - \check{x}_k|_K^2 + h_k$$

and compute the maximum in x_k, w_k of

$$-W_{k+1}(x_{k+1}) + W_k(x_k) + |x_k|_{Q-\gamma^2 C'N^{-1}C} + |\hat{u}_k|^2$$
$$+\gamma^2 x'_k C'N^{-1}y_k + \gamma^2 y'_k N^{-1}Cx_k - \gamma^2 |w_k|^2$$

under the constraint

$$Ax_k + B\widehat{u}_k + Dw_k = x_{k+1}.$$

Since we have already resolved the concavity issue, we need only write the stationarity conditions, which may be obtained with the use of a Lagrange multiplier, say $2p_k$, corresponding to the above constraint. Toward this end, introduce

$$\widehat{Q} := Q - \gamma^2 C'N^{-1}C, \quad \xi_k := C'N^{-1}y_k,$$

in terms of which the stationarity conditions read

$$\begin{aligned}w_k &= \gamma^{-2}D'p_k,\\ x_k &= (K+\widehat{Q})^{-1}(-K\check{x}_k + \gamma^2\xi_k + A'p_k).\end{aligned}$$

(These are in fact equivalent to (6.24)). The multiplier p_k is computed by placing this back into the constraint. Let

$$\begin{aligned}\check{x}_{k+1} &= A\check{x}_k - A(K+\widehat{Q})^{-1}(\widehat{Q}\check{x}_k + \gamma^2\xi_k) + B\widehat{u}_k\\ &\equiv A(K+\widehat{Q})^{-1}(K\check{x}_k - \gamma^2\xi_k) + B\widehat{u}_k,\end{aligned} \quad (6.49)$$

It is easily verified that the stationarity conditions may be written as

$$\begin{aligned}p_k &= -K(x_{k+1} - \check{x}_{k+1})\\ w_k &= \gamma^{-2}D'K(\check{x}_{k+1} - x_{k+1})\\ x_k &= (K+\widehat{Q})^{-1}[(K\check{x}_k - \gamma^2\xi_k) + A'K(x_{k+1} - \check{x}_{k+1})].\end{aligned}$$

This can be placed back into the function to be maximized, using the following form of (6.47):

$$K - KA(K+\widehat{Q})^{-1}A'K + \gamma^{-2}KDD'K = 0,$$

and we see that, provided that h_k is chosen properly (but independently of x_k), the maximum is zero as required. Now, (6.49) is the stationary version of (6.27) [Σ stands for Σ^+]:

$$\begin{aligned}\check{x}_{k+1} = A\check{x}_k + A(\Sigma^{-1} + C'N^{-1}C - \gamma^{-2}Q)^{-1}\\ \cdot[\gamma^{-2}Q\check{x}_k + C'N^{-1}(y_k - C\check{x}_k)] + B\widehat{u}_k\end{aligned} \quad (6.50)$$

exactly as (6.46) is the stationary version of (6.25a). The optimal controller is then the stationary version of (6.29), i.e., with M standing for M^+:

$$\widehat{u}_k = -B'(M + BB' - \gamma^{-2}DD')^{-1}A(I - \gamma^{-2}\Sigma M)^{-1}\check{x}_k. \tag{6.51}$$

The formulas (6.50)-(6.51) define the sequence $\{\widehat{u}_k\}$ as the output of a stationary linear system excited by $\{y_k\}$. It is therefore a trivial exercise to write this relationship in terms of a transfer function, although not very enlightening, since the formula is quite complicated.

We have thus proved the following result:

Theorem 6.4. *Consider the infinite-horizon discrete-time disturbance attenuation problem of this section, with (A, H) observable, (A, D) controllable, and with optimum attenuation level γ^*. If the algebraic Riccati equations (6.44) and (6.46) both admit (minimal) positive definite solutions M^+ and Σ^+, respectively, and if furthermore these matrices satisfy condition (6.48), then $\gamma \geq \gamma^*$, and a controller ensuring the attenuation level γ is given by (6.50)-(6.51). If one of the three conditions is violated, then $\gamma \leq \gamma^*$.* ◇

Example (continued). Let us continue with the illustrative example of Section 6.3, but this time with an infinite horizon. As seen in Section 3.4, we have

$$M^+ = \frac{1}{2}\left(1 + \sqrt{\frac{5\gamma^2 - 1}{\gamma^2 - 1}}\right),$$

and for the full state information problem, $\gamma^* = \sqrt{2}$. By the symmetry noticed earlier, we shall have

$$\Sigma^+ = M^+,$$

so that the global concavity condition leads to

$$\Sigma^+ M^+ < \gamma^2 \quad \Longleftrightarrow \quad M^+ < \gamma$$

which yields
$$\gamma^3 - \gamma^2 - 2\gamma + 1 > 0.$$

This is satisfied provided that
$$\gamma > \gamma^* \simeq 1.80194.$$

Again, the presence of disturbance in the measurement channel has increased the achievable attenuation level.

6.4 More General Classes of Problems

6.4.1 Cross terms in the performance index

The results presented heretofore in this chapter easily generalize to the case where the simplifying assumptions in (6.5) and (6.11) are replaced by

$$G'_k G_k =: R_k, \tag{6.53a}$$

$$H'_k G_k =: P_k, \tag{6.53b}$$

and

$$D_k E'_k =: L_k. \tag{6.54}$$

We have seen in Section 3.5.1 how a change of variable lets us deal with the cross term (6.53b). For the sake of completeness, and symmetry, we give here an alternative form of the equations yielding μ^*. The formulas being somewhat complicated, we introduce some simplifying notation. As in Section 5.5, let

$$\bar{A}_k = A_k - B_k R_k^{-1} P'_k, \quad \bar{Q}_k := Q_k - P_k R_k^{-1} P'_k \tag{6.55}$$

and let

$$\Gamma_{k+1} := M_{k+1}^{-1} + B_k R_k^{-1} B'_k - \gamma^{-2} D_k D'_k. \tag{6.56}$$

The Riccati equation (6.6) now reads

$$M_k = \bar{A}'_k \Gamma_{k+1}^{-1} \bar{A}_k + \bar{Q}_k, \quad M_{K+1} = Q_f \tag{6.57}$$

Imperfect State Measurements: Discrete Time

and the optimal full state feedback strategy for $\mathcal{P}_\gamma$ is

$$\mu_k^*(x_k) = -R_k^{-1}[B_k'\Gamma_{k+1}^{-1}\bar{A}_k + P_k']x_k. \tag{6.58}$$

Similarly, introduce

$$\widetilde{A}_k := A_k - L_k N_k^{-1} C_k, \quad \Pi_k := D_k D_k' - L_k N_k^{-1} L_k'. \tag{6.59}$$

We shall finally need the dual quantity to Γ:

$$\Delta_k := \Sigma_k^{-1} + C_k' N_k^{-1} C_k - \gamma^{-2} Q_k. \tag{6.60}$$

We may keep the cross terms in the derivation or use the same change of variables as in the proof of Theorem 6.3, but with the cross terms left in. This yields the Riccati equation

$$\Sigma_{k+1} = \bar{A}_k \Delta_k^{-1} \bar{A}_k' + \Pi_k, \quad \Sigma_1 = Q_0^{-1} \tag{6.61}$$

and the linear difference equation

$$\begin{aligned}\check{x}_{k+1} &= A_k \check{x}_k + B_k \widehat{u}_k + \gamma^{-2} \widetilde{A}_k \Delta_k^{-1} H_k' \check{z}_k \\ &\quad + (\widetilde{A}_k \Delta_k^{-1} C_k' + g_k) N_k^{-1}(y_k - C\check{x}_k), \end{aligned} \tag{6.62}$$

$$\check{x}_1 = 0$$

where, by definition,

$$\check{z}_k = H_k \check{x}_k + G_k \widehat{u}_k. \tag{6.63}$$

The relation (6.28), giving $\widehat{x}_k^k$ in terms of $\check{x}_k$, is unchanged, so that we obtain, by placing this in (6.58),

$$\widehat{u}_k = -R_k^{-1}(B_k'\Gamma_{k+1}^{-1}\bar{A}_k + P_k')(I - \gamma^{-2}\Sigma_k M_k)^{-1}\check{x}_k. \tag{6.64}$$

We thus arrive at the following complete result.

Theorem 6.5. *For the system (6.1)-(6.2) with the performance index given by (6.3b)-(6.4), and using the notation (6.5a), (6.53), (6.54), (6.55) and (6.59), suppose that the following three conditions hold over $[1, K]$:*

(i) The Riccati equation (6.57) has a solution satisfying

$$\Gamma_{k+1} - B_k' R_k^{-1} B_k > 0 \; ;$$

(ii) The Riccati equation (6.61) has a solution with $\Delta_k > 0$;

(iii) The global concavity condition (6.48) is satisfied.

Then, $\gamma \geq \gamma^*$, and a controller ensuring an attenuation level γ is given by (6.62) to (6.64). If any one of the three conditions above fails for some $k \in [1, K]$, then $\gamma \leq \gamma^*$. If the initial state is fixed at $x_1 = 0$, then the initialization of (6.61) must be taken as $\Sigma_2 = \Pi_1$, and the result holds provided that $\Pi_1 > 0$. For the stationary, infinite horizon problem, all equations must be replaced by their stationary counterparts, and (6.66) should be taken over $[0, \infty]$. Then, if the corresponding algebraic Riccati equations have (minimal) positive definite solutions satisfying (6.48), $\gamma \geq \gamma^*$; otherwise $\gamma \leq \gamma^*$. ◇

6.4.2 Delayed measurements

We consider here the case where, at time k, the available information is the sequence $y^{k-\theta}$ of the output $y_i, i = 1, \ldots, k - \theta$, where θ is a fixed integer. For $k \leq \theta$, no measurement is available. It is straightforward to extend the certainty-equivalence principle to this information structure, with the auxiliary problem at time τ being

$$\max_{\omega^{\tau-1} \in \Omega_{\tau-\theta}^{\tau-1}(\bar{u}^{\tau-2}, \bar{y}^{\tau-1})} G^{\tau-1}(\bar{u}^{\tau-1}, \omega^{\tau-1}). \tag{6.65}$$

(The definition of $\Omega_{\tau-\theta}^{\tau-1}$ is, of course, as in (6.15)). Noting that thus, w_k, $k = \tau - \theta + 1, \ldots, \tau - 1$, is unconstrained, we may extend Lemma 6.1:

$$\max_{\omega^\tau \in \Omega_{\tau-\theta}^\tau(\bar{u}, \bar{y})} G^\tau(\bar{u}^{\tau-\theta} \cdot \mu^*_{[\tau-\theta+1,\tau]}, \omega^\tau) = \max_{\omega^{\tau-\theta} \in \Omega_{\tau-\theta}^{\tau-\theta}} G^{\tau-\theta}(\bar{u}^{\tau-\theta}, \omega^{\tau-\theta}).$$

We conclude that the minimax control is obtained by solving the standard problem $\mathcal{P}_\gamma$ up to time $\tau - \theta$, and then taking $w_k = \nu_k^*(\widehat{x}_k^\tau)$ for $k = \tau - \theta$

up to $\tau - 1$, and using $u_\tau = \mu_\tau^*(\hat{x}_\tau^\tau)$ according to the certainty-equivalence principle. This is equivalent to applying the certainty-equivalence principle to the solution of the delayed state measurement problem of Section 3.2.3 (extended to an arbitrary finite delay).

Concavity of the auxiliary problem requires that the conditions of Theorem 6.3 on Σ be satisfied up to time $K - \theta$, and that the following maximization problem

$$\max_w \{V_{\tau+1}(x_{\tau+1}) + \sum_{k=\tau-\theta+1}^{\tau} g_k(x_k, \hat{u}_k, w_k)\} \tag{6.66}$$

with $\hat{u}_k$ fixed open-loop, be concave. The condition for this to be true is for the (open-loop) Riccati equation (see Lemma 3.1)

$$S_k = A_k' S_{k+1} A_k + A_k' S_{k+1} D_k (\gamma^{-2} I - D_k' S_{k+1} D_k)^{-1} D_k' S_{k+1} A_k + Q_k,$$

$$S_{\tau+1} = M_{\tau+1} \tag{6.67}$$

to have a solution over $[\tau - \theta + 1, \tau]$, that satisfies

$$\gamma^2 I - D_k' S_{k+1} D_k > 0. \tag{6.68}$$

The worst-case state $\hat{x}_\tau^\tau$ will now be obtained by solving

$$\hat{x}_{k+1}^\tau = (A_k - (B_k B_k' - \gamma^{-2} D_k D_k') \Gamma_{k+1}^{-1} A_k) \hat{x}_k^\tau,$$

using $\hat{x}_{\tau-\theta}^\tau$ obtained from (6.27)-(6.28) where the index τ must be replaced by $\tau - \theta$.

Theorem 6.6. *In the delayed information problem, with a fixed delay of length θ, let the following conditions be satisfied:*

(i) *The Riccati equation (6.6) has a solution over $[1, K+1]$, with*

$$M_{k+1}^{-1} - \gamma^{-2} D_k D_k' > 0;$$

(ii) *The set of Riccati equations (6.67) have solutions satisfying (6.68) over $[\tau - \theta + 1, \tau]$;*

(iii) The Riccati equation (6.25a) has a solution over $[1, K - \theta + 1]$, with $\Gamma_{k+1} > 0$;

(iv) Condition (6.30) is satisfied for $k \in [1, K - \theta + 1]$.

Then, $\gamma \geq \gamma^*$, and the procedure outlined prior to the statement of the theorem yields an optimal (minimax) controller. If any one of these four conditions fails, then $\gamma \leq \gamma^*$. ◇

6.4.3 The "filtering" problem

We assumed, in the standard problem $\mathcal{P}_\gamma$ that the information available at time k was y_i, $i = 1, \ldots, k-1$. Hence, we may view it as a one-step predictor problem. We consider here the case where in addition, y_k is available at the time u_k is computed.

Toward this end, we use the auxiliary criterion

$$G^\tau = |x_{\tau+1}|^2_{M_{\tau+1}} + |x_\tau|^2_{Q_\tau} + |u_\tau|^2 - \gamma^2|w_\tau|^2$$
$$+ \sum_{k=1}^{\tau-1}(|x_k|^2_{Q_k} + |u_k|^2 - \gamma^2|w_k|^2) - \gamma^2|x_1|^2_{Q_0}.$$

Using Lemma 6.1, we set

$$u_\tau = -B'_\tau(M^{-1}_{\tau+1} + B_\tau B'_\tau - \gamma^{-2}D_\tau D'_\tau)^{-1}A_\tau x_\tau =: -F_\tau x_\tau. \quad (6.69)$$

We may now compute, using dynamic programming, over one step

$$\max_{\omega^\tau \in \Omega_\tau} G^{\tau+1} = \max_{\omega^{\tau-1} \in \Omega_{\tau-1}} \left[\max_{w_\tau | y_\tau} G^\tau\right].$$

The maximum inside the bracket exists if

$$M^{-1}_{\tau+1} - \gamma^2 D_\tau D'_\tau > 0 \quad \text{or} \quad \Upsilon_{\tau+1} =: D'_\tau M_{\tau+1} D_\tau - \gamma^2 I < 0 \quad (6.70)$$

which is the standing assumption anyway, and the maximum value is easily found to be

$$-|y_\tau - C_\tau x_\tau|^2_{\gamma^2 N^{-1}_\tau} + |x_\tau|^2_{\Theta_\tau} + \sum_{k=1}^{\tau-1}(|x_k|^2_{Q_k} + |u_k|^2 - \gamma^2|w_k|^2) - \gamma^2|x_1|^2_{Q_0}$$

Imperfect State Measurements: Discrete Time 177

where

$$\Theta_\tau := A'_\tau(I - \Gamma_{\tau+1}^{-1} B_\tau B'_\tau)(M_{\tau+1} - M_{\tau+1} D_\tau \Upsilon_{\tau+1}^{-1} D'_\tau M_{\tau+1})$$
$$\cdot (I - B_\tau B'_\tau \Gamma_{\tau+1}^{-1}) A_\tau + F'_\tau F_\tau + Q_\tau. \tag{6.71}$$

(Notice that when DE' is not zero, the term in $y - Cx$ is more complicated).

We may use this new criterion in the theory as developed in Section 6.2. The only difference being in the final term, it will only show up in the final term of (6.24b) which is now replaced by

$$\lambda_\tau = \Theta_\tau x_\tau + \gamma^2 C'_\tau N_\tau^{-1} C_\tau (y_\tau - C_\tau x_\tau).$$

As a consequence, we have

$$\check{x}_\tau = \widehat{x}_\tau^\tau - \gamma^{-2} \Sigma_\tau \left[\Theta_\tau \widehat{x}_\tau^\tau + \gamma^2 C'_\tau N_\tau^{-1}(y_\tau - C \widehat{x}_\tau^\tau) \right],$$

and hence, (6.28) is replaced by

$$\widehat{x}_\tau^\tau = \left[I - \gamma^{-2} \Sigma_\tau (\Theta_\tau - \gamma^2 C'_\tau N_\tau^{-1} C_\tau) \right]^{-1} (\check{x}_\tau + \Sigma_\tau C'_\tau N_\tau^{-1} y_\tau) \tag{6.72}$$

while the global concavity condition (6.30) becomes

$$\rho[\Sigma_\tau (\Theta_\tau - \gamma^2 C'_\tau N_\tau^{-1} C_\tau)] < \gamma^2. \tag{6.73}$$

We have therefore obtained the following result:

Theorem 6.7. *For the discrete-time disturbance attenuation problem of Section 6.1, assume that also the current value of the output is available for the controller, and for this problem let γ^* again denote the optimum attenuation level. Let the conditions of Theorem 6.4, with (6.30) replaced by (6.73), stand. When $\gamma \geq \gamma^*$, the solution is obtained as in Theorem 6.3, except that (6.28) is now replaced by (6.72) and (6.29), which is $u_k = \mu_k^+(\widehat{x}_k^k)$, modified accordingly. (The notation used in (6.72) and (6.73) is defined in (6.69) to (6.71)).* ◊

6.4.4 Nonlinear/nonquadratic problems

Consider the nonlinear problem of Section 3.6, but with an imperfect nonlinear measurement as in (6.13b). We may keep the quadratic measure of noise intensity and of initial state amplitude, or change them to any other positive measure. Assume that the soft-constrained dynamic game associated with this problem has been solved under the CLPS information pattern, leading to a value function $V_k(x)$ and a feedback minimax strategy $\mu_k^*(x)$. We have already pointed out in Remark 6.1 that the certainty equivalence principle (i.e., the results of Theorems 6.1 and 6.2) holds for the nonlinear problem as well, provided that the various maxima involved exist and are unique. Therefore, we may still define an auxiliary problem as in (6.15) to (6.17), and employ Theorem 6.1.

Notice that the computation of the value function V can be done offline. Then, the forward dynamic programming approach as in (6.33) with the constraints set to

$$x_{k+1} = f_k(x, u_k, w) \quad \text{and} \quad h_k(x, w) = y_k$$

can in principle be carried out recursively in real time, yielding W_{k+1} when y_k becomes available. Therefore,

$$\hat{x}_{k+1}^{k+1} = \arg\max_x [V_{k+1}(x) + W_{k+1}(x)]$$

which can also, in principle, be computed at that time, and placed into μ_{k+1}^* to yield the minimax controller, $\mu_{k+1}^*(\hat{x}_{k+1}^{k+1})$.

6.5 Main Points of the Chapter

This chapter has developed the discrete-time counterparts of the continuous-time results of Chapter 5, by establishing again a certainty-equivalence type of decomposition. For linear systems where the controller has access to the entire past values of the disturbance-corrupted state measurements, the characterization of a controller that guarantees a given disturbance attenuation level is again determined by the solutions of two (discrete-time) Riccati equations ((6.6) and (6.25a)), and a spectral radius condition (6.30b). As can be seen from (6.27)-(6.29), the controller is a certainty-equivalence controller, with the compensator (or estimator) having the same dimension as that of the plant. A precise statement on the optimality of such a controller has been given in Theorem 6.3, whose counterpart in the "filtering" case where the controller has access to also the current value of the measurement vector is Theorem 6.7. The chapter has also presented results on the infinite-horizon case (see Theorem 6.4), on problems with more general cost functions with cross terms in the state and control (cf. Theorem 6.5), and delayed information structures (cf. Theorem 6.6). Furthermore, some discussion on the nonlinear/nonquadratic problem has been included in the last subsection of the chapter.

The solution to the basic problem covered by Theorem 6.3 was first outlined in [23], where complete proofs were not given. Some early work on this problem has also been reported in [60]. The form in which the results are established here, and the solutions to the more general problems discussed, also involve the contributions of Garry Didinsky, a Ph.D. student at UI. As far as we know, this is the first published account of these results, which completes the theory of H^∞-optimal control for linear-quadratic systems, in both continuous and discrete time, and under arbitrary information structures.

Chapter 7

Performance Levels For Minimax Estimators

7.1 Introduction

In the previous two chapters, in the development of the solutions to the disturbance attenuation problem with imperfect measurements in continuous and discrete time, we have encountered filter equations which resemble the standard Kalman filter when the weighting on the state in the cost function (i.e., Q) is set equal to zero. In this chapter we study such problems, independently of the analysis of the previous chapters, and show that the appearance of the Kalman filter is actually not a coincidence. The commonality of the analysis of this chapter with those of the previous ones is again the use of game-theoretic techniques.

To introduce the class of estimation problems covered here, let us first consider the discrete-time formulation. Suppose that we are given the linear time-varying system

$$x_{k+1} = A_k x_k + D_k w_k, \quad k \geq 0; \quad x_k \in \mathbb{R}^n, w_k \in \mathbb{R}^m \qquad (7.1)$$

along with the linear measurement equation

$$y_k = C_k x_k + v_k; \quad v_k \in \mathbb{R}^p, \qquad (7.2)$$

and are interested in obtaining an estimate, δ_K, for the system state at time $k = K$, using measurements collected up to (and including) that time,

$y_{[0,K]}$. The error in this estimate is

$$e_K := x_K - \delta_K(y_{[0,K]}), \qquad (7.3)$$

which depends on the system and measurement noises ($\{w_k, k \geq 0\}$ and $\{v_k, k \geq 0\}$, respectively), as well as on the initial state x_0. The *Bayesian* approach to this (*filtering*) problem assigns probability distributions to the unknowns, and determines δ_K as the function that minimizes the expected value of a particular norm of e_K. For a large class of such norms, the unique optimum is the conditional mean of x_K, which is well-known to be linear if the random vectors are also jointly Gaussian distributed. Moreover if the noise sequences are independent white noise sequences, then the conditional mean is generated recursively by the Kalman filter [1].

The alternative (*minimax*) approach to be discussed here, however, is purely deterministic. Toward formulating this class of problems, let $\|w, v, x_0\|_K$ denote some Euclidean norm of the $(2K+2)$-tuple $(w_{[0,K-1]}, v_{[0,K]}, x_0)$, and introduce the mapping

$$T_{\delta_K}(w, v, x_0) := x_K - \delta_K(y_{[0,K]}) \equiv e_K \qquad (7.4)$$

which is the counterpart of the mapping T_μ of Section 1.2. Then, consistent with (1.2), we will call an estimator δ_K^* *minimax* if the following holds:

$$\begin{aligned}\inf_{\delta_K} \sup_{w,v,x_0} & \left\{ |T_{\delta_K}(w,v,x_0)| / \|w,v,x_0\|_K \right\} \\ = \sup_{w,v,x_0} & \left\{ |T_{\delta_K^*}(w,v,x_0)| / \|w,v,x_0\|_K \right\} =: \gamma_K^*.\end{aligned} \qquad (7.5)$$

Here γ_K^* is the minimax attenuation level for this K stage filtering problem.

In the above formulation, if the measurements available for the estimation of x_K are $y_{[0,K-\ell]}$, then this constitutes the ℓ-step *prediction* problem; if, on the other hand, we are interested in the estimation of x_ℓ, for some $\ell < K$, using measurements $y_{[0,K]}$, then the problem is one of *smoothing*, in which case we could also replace (7.4) by the mapping whose norm is

$$\|T_{\delta_{[0,K]}}(w,v,x_0)\| := \left\{ \sum_{k=0}^{K} |x_k - \delta_k(y_{[0,K]})|_{N_k}^2 \right\}^{\frac{1}{2}} \qquad (7.6)$$

where $N_k \geq 0$ is some weighting matrix.

In the continuous-time version of these three classes of problems, the state equation (7.1) is replaced by the differential equation

$$\dot{x} = A(t)x + D(t)w(t), \qquad (7.7)$$

and (7.2) is replaced by the continuous-time equation

$$y(t) = C(t)x(t) + v(t), \qquad (7.8)$$

where $w \in \mathcal{H}_w$ and $v \in \mathcal{H}_v$ are square-integrable functions on the interval $[0, t_f]$. The performance index is the same as in the discrete-time case, with appropriate choice of norms.

In the next section, we discuss the solution of a particular linear-quadratic game, which plays an important role in the solution of these problems. The optimum performance bound for each such problem is then derived in Section 7.3. Section 7.4 provides some concluding remarks.

Note that the estimation problems identified above involve pointwise (terminal state) error, and not cumulative error. If we have to make a comparison with the formulation of Chapter 1, this corresponds to the case where the objective function contains only a terminal penalty (i.e., Q_f) on the state; of course, in addition to this, the systems here are control-free. In the above formulation, if we also include a cumulative error in the performance index, the resulting minimax estimation problem is known as an H^∞-filtering (or prediction) problem, which has been studied recently in [52].

7.2 A Static Minimax Estimation Problem

Let x, w, v be finite-dimensional vectors, and two other vectors, z and y, be defined by

$$z = Mx + Dw \qquad (\in \mathbb{R}^n) \qquad (7.9)$$

$$y = Hx + Fw + v \qquad (\in \mathbb{R}^p) \qquad (7.10)$$

where M, D, H, F are matrices of appropriate dimensions. In order to avoid some trivial cases, we will assume that the matrix $(M,D)'(M,D)$ has at least one positive eigenvalue (*i.e.* it is not the *zero* matrix). Let

$$\begin{aligned} \|w,v,x\|^2 &:= x'x + v'Rv + w'w; \quad R > 0 \\ &\equiv |x|^2 + |v|_R^2 + |w|^2 \end{aligned} \quad (7.11)$$

and

$$T_\delta(w,v,x) = z - \delta(y). \quad (7.12)$$

Let Δ be the class of all Borel measurable mappings $\delta : \mathbb{R}^p \to \mathbb{R}^n$. Then, we seek a $\delta^* \in \Delta$, and a $\gamma^* \in \mathbb{R}$, such that (parallel to (7.5))

$$\begin{aligned} &\sup_{w,v,x} \{|T_{\delta^*}(w,v,x)|/\|w,v,x\|\} \\ &= \inf_{\delta \in \Delta} \sup_{w,v,x} \{|T_\delta(w,v,x)|/\|w,v,x\|\} =: \gamma^*. \end{aligned} \quad (7.13)$$

Toward this end, we first introduce a related quadratic game with a kernel parameterized by γ:

$$L_\gamma(\delta(y); x, w, v) = |Mx + Dw - \delta(y)|^2 - \gamma^2 \|w,v,x\|^2 \quad (7.14)$$

for which we seek the *upper value* with δ being the minimizer, and x, w, v the maximizer. More precisely, we seek a $\bar{\delta}_\gamma \in \Delta$ such that

$$\sup_{w,v,x} L_\gamma(\bar{\delta}_\gamma(y); x, w, v) = \inf_{\delta \in \Delta} \sup_{w,v,x} L_\gamma(\delta(y); x, w, v) =: \bar{L}_\gamma. \quad (7.15)$$

The following Lemma provides the complete solution to this quadratic game:

Lemma 7.1. *For the quadratic game defined by kernel (7.14),*

i) *There exists a $\hat{\gamma} > 0$ such that for $\gamma \geq \hat{\gamma}$ the upper value $\bar{L}_\gamma$ is finite (and actually zero), whereas for $0 \leq \gamma < \hat{\gamma}$, it is infinite.*

ii) *For all $\gamma \geq \hat{\gamma}$, the game admits a minimax controller given by*

$$\bar{\delta}_\gamma(y) = \bar{\delta}(y) = (MH' + DF')\left(R^{-1} + HH' + FF'\right)^{-1} y \quad (7.16)$$

iii) $(\hat{\gamma})^2$ is the smallest positive scalar r for which

$$\sup_{x,w} \{|\, Mx + Dw\,|^2 - r\left(|x|^2 + |Hx + Fw|_R^2 + |w|^2\right)\} = 0, \quad (7.17)$$

or equivalently it is the maximum eigenvalue (i.e., the spectral radius) of the matrix

$$\Lambda := [I + (H,F)'R(H,F)]^{-\frac{1}{2}} (M,D)'(M,D) [I + (H,F)'R(H,F)]^{-\frac{1}{2}} \quad (7.18)$$

where $(\cdot)^{-\frac{1}{2}}$ denotes the inverse of the unique positive definite square root of the matrix $(\cdot)$.

Proof. The proof follows through the line of reasoning used in the derivation of dynamic Stackelberg strategies [4], and is also in line with the arguments used in deriving the main results of Chapters 5 and 6. Since δ is a (Borel) function of y,

$$\begin{aligned}
\bar{L}_\gamma &= \inf_{\delta \in \Delta} \sup_{x,w,y} L_\gamma(\delta(y); x, w, y - Hx - Fw) \\
&= \sup_y \inf_u \sup_{w,x} L_\gamma(u; x, w, y - Hx - Fw)
\end{aligned} \quad (\circ)$$

which has converted the original dynamic optimization problem into an equivalent static one. Now, for each fixed $y \in {\rm I\!R}^p$, we study the inner "inf sup" problem, which involves a quadratic kernel. This kernel is strictly convex in u but not necessarily strictly concave in the pair (x, w), with the latter condition holding if, and only if,

$$\Lambda - \gamma^2 I < 0. \quad (*)$$

We know from the theory of zero-sum quadratic games ([[17], p. 186]; see also Theorem 2.3) that under condition $(*)$ the game admits a unique saddle-point solution, whereas if the matrix in $(*)$ has at least one positive eigenvalue, concavity (in (x,w)) is lost, implying that the upper value is unbounded.

Now, if (∗) holds (i.e., a saddle point exists to the inner game in (◦)),

$$\inf_{u} \sup_{x,w} L_\gamma = \max_{x,w} \min_{u} L_\gamma = \min_{x,w} \|w, y - Hx - Fw, x\|^2$$

$$\equiv \|\widehat{w}, y - H\widehat{x} - F\widehat{w}, \widehat{x}\|^2$$

where

$$\widehat{w}(y) = F'(R^{-1} + HH' + FF')^{-1} y$$

$$\widehat{x}(y) = H'(R^{-1} + HH' + FF')^{-1} y.$$

Furthermore, since the minimizing u was

$$u = Mx + Dw,$$

(7.16) readily follows as a minimax policy for $\gamma > \hat{\gamma}$. Note that for $\gamma > \hat{\gamma}$, the minimax policy is independent of γ, and furthermore

$$\sup_{y} \inf_{u} \sup_{x,w} L_\gamma = 0.$$

We now claim that (by continuity) the same holds for $\gamma = \hat{\gamma}$, and hence condition (∗) in the limit is equivalent to (7.17). This is true because, as just proven,

$$\hat{\gamma} \leq \sup_{w,v,x} \{|T_{\bar{\delta}}(w,v,x)|/\|w,v,x\|\} < \gamma$$

for all $\gamma > \hat{\gamma}$. Note that $\hat{\gamma} > 0$ since we had taken $(M, D)'(M, D) \neq 0$. This completes the proof of all parts of the Lemma. ◇

We are now in a position to present the solution to (7.13).

Theorem 7.1. *The minimax estimator for the static problem of this section is* $\delta^* = \bar{\delta}$, *which is defined by (7.16). The minimax attenuation level* γ^* *is equal to* $\hat{\gamma}$, *which is the square root of the maximum eigenvalue of (7.18).*

Proof. In view of Lemma 7.1, we have for all (x, w, v),

$$L_{\hat{\gamma}}(\bar{\delta}(y); x, w, v) \leq 0$$

which implies

$$|T_{\bar{\delta}}(w, v, x)| \leq \hat{\gamma} \|w, v, x\|,$$

and since $\hat{\gamma}$ is the smallest such scalar, (7.13) follows. ◇

We now conclude this section with a crucial observation.

Corollary 7.1. *The minimax estimator, δ^*, of Theorem 7.1 is a Bayes estimator for z under the measurement y, when x, w, v are zero-mean independent Gaussian random vectors with respective covariances I, I, and R^{-1}. Equivalently, under this a priori distribution, δ^* is the conditional mean:*

$$\delta^*(y) = E[z \mid y] \tag{7.19}$$

Proof. This follows from a well-known property of Gaussian distributions [1]. ◇

Remark 7.1. The above result should not be construed as saying that the Gaussian distribution is least favorable [38] for a game with kernel as in (7.13); it is simply an interpretation given to γ^*, which will prove to be very useful in the next section. ◇

Remark 7.2. Lemma 7.1 as well as Theorem 7.1 extend naturally to the case where x, w, v, y and z belong to infinite-dimensional complete inner product (Hilbert) spaces. In this case M, D, H and F will be taken as bounded linear operators, and R will be taken as a strongly positive linear operator. The Euclidean norms will be replaced by the norms on the corresponding Hilbert spaces, and the scalar products by inner products. Then, the counterpart of (7.16) will be

$$\bar{\delta}(y) = (MH^* + DF^*)(R^{-1} + HH^* + FF^*)^{-1}y, \tag{7.20}$$

where "*" denotes the adjoint of a linear operator. Condition (7.18) is also equally valid here, with "'" replaced by "*". ◇

7.3 Optimum Performance Levels

7.3.1 Discrete time

We now return to the dynamic estimation problems formulated in Section 7.1, where we take the disturbance norm as

$$\|w, v, x_0\|_K^2 = |x_0|^2 + \sum_{k=0}^{K-1} \{|w_k|^2 + |v_k|_{R_k}^2\} + |v_K|_{R_K}^2 \tag{7.21}$$

where R_k is a positive definite weighting coefficient matrix, for every $k = 0, 1, \ldots, K$. Note that there is no loss of generality in taking the norms on x_0 and w_k as standard Euclidean norms, because any nonunity weighting can be absorbed into the problem parameters, by redefining x_0 and w_k.

a) Filtering

Since we have a terminal state estimation problem (see (7.4)), this problem is no different from the static one formulated in the previous section, and hence Theorem 7.1 equally applies here. To see the direct correspondence, let us rewrite x_k in the form (with some obvious definitions for Φ_k and N_k):

$$x_k = \Phi_k x_0 + N_k w_{[0,k-1]} \tag{7.22a}$$

where $w_{[0,k-1]}$ is a column vector, made up of the w_ℓ's, $\ell = 0, \ldots, k-1$. Then $y_{[0,K]}$ can be rewritten as

$$y_{[0,K]} = H x_0 + F w_{[0,K-1]} + v_{[0,K]} \tag{7.22b}$$

for some H and F, where F is lower block triangular. Now letting $M := \Phi_K$, $D := N_K$, the problem becomes equivalent to the one of the previous section, and hence by Corollary 7.1,

$$\delta_K^* \left(y_{[0,K]}\right) = E\left[x_K \mid y_{[0,K]}\right] \tag{7.23}$$

with $\{w_k\}$, $\{v_k\}$ being independent zero-mean Gaussian white noise sequences, where $w_k \sim N(0, I)$, $v_k \sim N(0, R_k^{-1})$. Furthermore $x_0 \sim N(0, I)$, and it is independent of $\{w_k\}$ and $\{v_k\}$. Consequently, δ_K^* is generated by a Kalman filter. In fact, the same Kalman filter generates the minimax estimate for the filtering problem defined on any subinterval of $[0, K]$ and using the same parameter values.

Now, to determine γ^*, we have to obtain the counterpart of the optimization problem (7.17). Using the correspondence above, the equivalent optimization problem can easily be seen to be

$$\max_{w, x_0} \left\{ |x_K|^2 - \gamma^2 \left\{ |x_0|^2 + \sum_{k=0}^{K-1} |w_k|^2 + \sum_{k=0}^{K} |C_k x_k|_{R_k}^2 \right\} \right\}$$

$\Leftrightarrow$

$$\min_{w, x_0} \left\{ x_K' \left(C_K' R_K C_K - \frac{1}{\gamma^2} I \right) x_K + |x_0|^2 \right. \tag{7.24}$$
$$\left. + \sum_{k=0}^{K-1} (x_k' C_k' R_k C_k x_k + w_k' w_k) \right\}$$

subject to the system equation

$$x_{k+1} = A_k x_k + D_k w_k.$$

We seek the smallest value of γ^2 under which this Linear-Quadratic (LQ) control problem has a finite (equivalently *zero*) cost. Using Lemma 3.1, we now arrive at the following result:

Theorem 7.2. *Let Γ_F be the set of all $\gamma > 0$ for which the sequence of symmetric matrices $\{S_k\}_{k=K}^{0}$ generated by the Riccati equation*

$$S_k = C_k' R_k C_k + A_k' S_{k+1} \left[I - D_k P_k^{-1} D_k' S_{k+1} \right] A_k; \tag{7.25a}$$
$$S_K = C_K' R_K C_K - \gamma^{-2} I$$

$$P_k := I + D_k' S_{k+1} D_k \tag{7.25b}$$

satisfy the following two conditions:

i) P_k, $k = 0, \ldots, K-1$, are positive definite

ii) $S_0 + I$ is nonnegative definite

Then, $\gamma_F := \inf\{\gamma : \gamma \in \Gamma_F\}$ is the minimax attenuation level γ_K^* for the filtering problem. ◇

b) Prediction

By the same reasoning as in the filtering problem, the minimax ℓ-step predictor is

$$\delta_K^*(y_{[0,K-\ell]}) = E[x_K \mid y_{[0,K-\ell]}]$$

where the statistics on the disturbances are as before. Hence δ^* is a Kalman predictor.

The counterpart of the minimization problem (7.24) in this case is

$$\min_{w,x_0} \left\{ -\frac{1}{\gamma^2} |x_K|^2 + |x_0|^2 + \sum_{k=0}^{K-\ell} x_k' C_k' R_k C_k x_k + \sum_{k=0}^{K-1} w_k' w_k \right\}. \quad (7.26)$$

Hence, the result of Theorem 7.2 equally applies here, provided that we take $R_k = 0$ for $k > K - \ell$. Denoting the minimax attenuation level in this case by γ_P, clearly we have $\gamma_P \geq \gamma_F$.

c) Smoothing

Here Theorem 7.1 and Corollary 7.1 are directly applicable, since there is no causality requirement on the estimator. This also makes the resulting estimator an H_∞-optimal smoother. Using the notation of (7.6), we have

$$\delta_k^*(y_{[0,K]}) = E[x_k \mid y_{[0,K]}], \qquad k = 0, 1, \ldots, K$$

where the distributions are again as before. Hence, the standard Bayes smoothers (involving two recursive equations -- one in forward and one in backward time [1]) are also minimax estimators. The condition (7.17) now becomes

$$\min_{w,x_0} \left\{ -\frac{1}{\gamma^2} \sum_{k=0}^{K} |x_k|_{N_k}^2 + |x_0|^2 + \sum_{k=0}^{K} x_k' C_k' R_k C_k x_k + \sum_{k=0}^{K-1} w_k' w_k \right\} = 0, \quad (7.27)$$

under the system equation (7.1).

The following counterpart of Theorem 7.2 readily follows for the smoothing problem.

Theorem 7.3. *Let Γ_S be the set of all $\gamma > 0$ for which the sequence of symmetric matrices $\{S_k\}_{k=K}^0$ generated by the Riccati equation*

$$S_k = C_k' R_k C_k - \frac{1}{\gamma^2} N_k + A_k' S_{k+1} \left[I - D_k P_k^{-1} D_k' S_{k+1} \right] A_k; \qquad (7.28a)$$

$$S_K = C_K' R_K C_K - \frac{1}{\gamma^2} N_K$$

$$P_k := I + D_k' S_{k+1} D_k \qquad (7.28b)$$

satisfy the following two conditions:

i) P_k, $k = 0, \ldots, K-1$, are positive definite

ii) $S_0 + I$ is nonnegative definite.

Then, $\gamma_S := \inf\{\gamma : \gamma \in \Gamma_S\}$ is the minimax attenuation level for the smoothing problem. ◇

7.3.2 Continuous time

For the continuous-time problem, we take as the counterpart of (7.21):

$$\|w, v, x_0\|^2 = |x_0|^2 + \int_0^{t_f} \left\{ |w(t)|^2 + |v(t)|_{R(t)}^2 \right\} dt. \qquad (7.29)$$

Then it is a simple exercise to write down the counterparts of the three minimization problems (7.24), (7.26) and (7.27), and the associated (continuous-time) Riccati equations. We assume below that all coefficient matrices have piecewise continuous entries on $[0, t_f]$.

a) **Filtering**

The optimization problem is:

$$\min_{w, x_0} \left\{ -\frac{1}{\gamma^2} |x(t_f)|^2 + |x_0|^2 + \int_0^{t_f} [x(t)' C' R C x(t) + w(t)' w(t)] dt \right\} \qquad (7.30)$$

subject to (7.8).

Then, the following theorem follows from Lemma 4.1:

Theorem 7.4. *Let Γ_F^c be the set of all $\gamma > 0$ for which the Riccati equation below does not have any conjugate point in the interval $[0, t_f]$:*

$$\dot{S} + SA + A'S - SDD'S + C'RC = 0, \quad S(t_f) = -\frac{1}{\gamma^2}I, \quad (7.31)$$

and furthermore that

$$S(0) + I \geq 0.$$

Then, $\gamma_F^c := \inf\{\gamma : \gamma \in \Gamma_F^c\}$ is the minimax attenuation level for the filtering problem.

◊

b) **Prediction**

If we allow a θ unit delay in the utilization of the measurement, (7.30) is replaced by

$$\min_{w,x_0} \left\{ -\frac{1}{\gamma^2}|x(t_f)|^2 + |x_0|^2 + \int_0^{t_f - \theta} x(t)'C'RCx(t)dt + \int_0^{t_f} |w(t)|^2 dt \right\}. \quad (7.32)$$

Hence, to obtain the minimax attenuation level, γ_P^c, we simply have to take $R(t) \equiv 0$, for $t > t_f - \theta$, in the statement of Theorem 7.4.

c) **Smoothing**

If we take the integrated smoothing error as

$$\int_0^{t_f} |x(t) - \delta_t(y(s); 0 \leq s < t_f)|_{N(t)}^2 dt, \quad (7.33)$$

the associated optimization problem becomes

$$\min_{w,x_0} \left\{ |x_0|^2 + \int_0^{t_f} \left[x(t)'[C'RC - \frac{1}{\gamma^2}N(t)]x(t) + w(t)'w(t) \right] dt \right\} \quad (7.34)$$

for which the counterpart of Theorem 7.4 is the following.

Theorem 7.5. *Let Γ_S^c be the set of all $\gamma > 0$ for which the Riccati differential equation below does not have any conjugate point in the interval $[0, t_f]$:*

$$\dot{S} + SA + A'S - SDD'S + C'RC - \frac{1}{\gamma^2}N = 0; \quad S(t_f) = 0, \qquad (7.35)$$

and furthermore that

$$S(0) + I \geq 0.$$

Then, $\gamma_S^c := \inf\{\gamma : \gamma \in \Gamma_F^c\}$ is the minimax (H^∞) attenuation level for the smoothing problem. ◇

In all cases above the infinite-dimensional versions of Theorem 7.1 and Corollary 7.1 (à la Remark 7.2) can be used to establish the Bayes property of the minimax estimators. The corresponding distributions for the disturbances are again Gaussian, with $w(t)$, $v(t)$ being independent zero-mean white noises (more precisely, their integrals are Wiener processes), with the associated covariances being $I\delta(t)$ and $R(t)\delta(t)$, respectively, where $\delta(\cdot)$ is the Dirac delta function. Hence, again the well-known Bayesian filter, predictor and smoother equations generate the minimax estimators.

We should also note that if sampled (instead of continuous) measurements are available, the Riccati equations above will have to be modified to accommodate also this type of measurement schemes; this should now be an easy exercise, given the analyses of the previous chapters on the sampled-data information pattern.. Furthermore, if the initial state x_0 is instead a known quantity (say the *zero* vector), then there would not be any need for the second conditions in the theorems above (i.e., those that involve $S(0)$).

7.4 Summary of Main Results

This chapter has provided a nutshell analysis of a class of minimax filter, prediction and smoothing problems for linear time-varying systems in both discrete and continuous time, by making use of the saddle point of a particular quadratic game. The main structural difference between the performance index here and those adopted in the previous chapters is that here the "gain" is from the disturbance(s) to the *pointwise* (instead of cumulative) output (which in this case is the estimation error). This difference in the cost functions leads to an important structural difference in the solutions, in that for the class of problems studied here the minimax decision rules (estimators) can be obtained without computing the associated minimax attenuation levels – – a feature the decision rules (controllers) obtained in the previous chapters did not have. As a result, here the minimax estimators and the associated performance levels can be determined independently, with the former being Bayes estimators with respect to Gaussian distributions (which then readily leads to the Kalman-type recursive structures), and the latter (that is, the minimax disturbance attenuation levels) determined from the solutions of some related linear quadratic (indefinite) optimal control problems, as presented in Theorems 7.2-7.5.

The derivation here is based on the recent work reported in [12], though the fact that these minimax estimators are Bayes estimators with respect to Gaussian distributions had in fact been presented some twenty years ago in [28]. The authors' primary interest in [28] was to obtain a complete characterization of the evolution of the uncertainty in the value of the state x_k (or $x(t)$), caused by the norm (or energy) bounded disturbances w, v and x_0, and in the light of the measurements received. Indeed, it was shown in [28] that these uncertainty sets are ellipsoids, whose centroids (which are minimax estimators) are generated by Kalman filter (predictor or smoother) type equations, and independently of the magnitudes of the norm bounds.

Chapter 8

Appendix A: Conjugate Points

We provide, in this appendix, a self-contained introduction to *conjugate points* as they arise in dynamic linear-quadratic optimization, and in particular in the context of linear-quadratic differential games. The results presented below are used in Chapter 4, in the proof of some key results in the solution of the continuous-time disturbance attenuation problem.

We consider a two-player system in $I\!R^n$, defined over a time interval $[0, t_f]$:

$$\dot{x}(t) = A(t)x(t) + B(t)u(t) + D(t)w(t), \qquad x(0) = x_0. \qquad (8.1)$$

The matrix functions $A(\cdot)$, $B(\cdot)$ and $D(\cdot)$ are assumed to be piecewise continuous and bounded over $[0, t_f]$, and have dimensions $n \times n$, $n \times m_1$ and $n \times m_2$, respectively. We are also given a nonnegative definite matrix Q_f and a nonnegative definite matrix function $Q(\cdot)$, both of dimension $n \times n$. We let

$$|x|^2_{Q_f} := x'Q_f x \quad \text{and} \quad \|x\|^2_Q := \int_0^{t_f} x'(t)Q(t)x(t)dt.$$

Let $\mathcal{U} = L^2([0, t_f], I\!R^{m_1})$, $\mathcal{W} = L^2([0, t_f], I\!R^{m_2})$, $\|u\|$ and $\|w\|$ denote the L^2 norms of $u \in \mathcal{U}$ and $w \in \mathcal{W}$. For every positive $\gamma \in I\!R^+$, let $J_\gamma : I\!R^n \times \mathcal{U} \times \mathcal{W} \to I\!R$ be given by

$$J_\gamma(x_0; u, w) = |x(t_f)|^2_{Q_f} + \|x\|^2_Q + \|u\|^2 - \gamma^2 \|w\|^2,$$

where $t \mapsto x(t)$ is the unique solution of (8.1). When there is no source of ambiguity, we shall suppress the subindex γ of J.

We first consider the minimization of J with respect to u, for fixed $w \in \mathcal{W}$. The function $u \mapsto x$ is affine, and $(u, x) \mapsto J$ is quadratic, and therefore $u \mapsto J(x_0; u, w)$ is quadratic nonhomogeneous. Since $Q_f \geq 0$, and $Q \geq 0$, it follows that J is convex, and $J(x_0, u, w) \to \infty$ as $\|u\| \to \infty$ (the reason here being that $J(x_0; u, w) \geq \|u\|^2 - \gamma^2 \|w\|^2$). Therefore J has a (unique) minimum, which can be obtained using the standard necessary conditions of the calculus of variations, or the Pontryagin minimum principle, as

$$u(t) = -B'(t)\lambda(t) \tag{8.2a}$$

where $\lambda(t)$ is given, along with the optimal trajectory, as the solution of the Hamiltonian system (a two point boundary value problem)

$$\dot{x} = Ax - BB'\lambda + Dw, \qquad x(0) = x_0 \tag{8.2b}$$

$$\dot{\lambda} = -Qx - A'\lambda, \qquad \lambda(t_f) = Q_f x(t_f). \tag{8.2c}$$

It follows from an analysis identical to the one to be developed shortly for the maximization problem that there exists a solution $K(t)$ to the associated Riccati differential equation (RDE)

$$\dot{K} + KA + A'K - KBB'K + Q = 0, \qquad K(t_f) = Q_f, \tag{8.3}$$

and that the optimal control and optimal cost can be written in terms of K and the associated variables.

Let us now turn to the maximization problem. We first prove two theorems (namely, Theorems 8.1 and 8.2, to follow) that can be obtained by elementary means. We shall then state and prove a more complete result, given as Theorem 8.3.

Introduce the following Riccati differential equation for a matrix S (the "maximization Riccati equation")

$$\dot{S} + SA + A'S + \gamma^{-2}SDD'S + Q = 0, \qquad S(t_f) = Q_f. \tag{8.4}$$

Whenever necessary to indicate the explicit dependence on the parameter γ, we shall denote the solution of this equation by $S_\gamma(\cdot)$. By classical analysis, (8.4) has a solution in a (left) neighborhood of the terminal time t_f. However, it may have a finite escape time in $[0, t_f)$, as the following classical example shows:

Given the RDE

$$\dot{S} + S^2 + 1 = 0; \qquad t_f = \frac{\pi}{2}; \qquad S\left(\frac{\pi}{2}\right) = 1,$$

solving it in retrograde time we obtain

$$S(t) = \tan\left(\frac{3\pi}{4} - t\right)$$

which diverges at $t = \frac{\pi}{4}$.

Definition 8.1. *A finite escape time of (8.4) is a conjugate point.* ◇

We have the following important result.

Proposition 8.1. *Whenever it exists, $S_\gamma(t)$ is nonnegative definite. If either Q_f or $Q(\cdot)$ is positive definite, so is $S_\gamma(t)$.*

Proof. Since by hypothesis $S_\gamma(\tau)$ exits for $\tau \in [t, t_f]$, on the same interval we can introduce the transition matrix $\Psi(\cdot, \cdot)$ associated with $A + (\gamma^{-2}/2)DD'S_\gamma$. Note that (8.4) can be rewritten as

$$\dot{S} + S\left(A + \frac{\gamma^{-2}}{2}DD'S\right) + \left(A' + \frac{\gamma^{-2}}{2}SDD'\right)S + Q = 0; \qquad S(t_f) = Q_f,$$

so that $S_\gamma(t)$ admits the following representation

$$S_\gamma(t) = \Psi'(t_f, t)Q_f\Psi(t_f, t) + \int_t^{t_f} \Psi'(\tau, t)Q(\tau)\Psi(\tau, t)d\tau$$

from which the result follows, since Ψ is always invertible. ◇

Theorem 8.1. *If (8.4) admits a solution defined over $[0, t_f]$, then*

$$\sup_w J_\gamma(x_0; u, w) < \infty.$$

Appendix A: Conjugate Points

Proof. We use the classical device of *completing the squares*, but for a nonhomogeneous problem. Introduce two additional differential equations, for a vector function $p(t) \in \mathbb{R}^n$ and a scalar function $q(t)$:

$$\dot{p} + (A' + \gamma^{-2}SDD')p + SBu = 0, \qquad p(t_f) = 0, \qquad (8.5a)$$

$$\dot{q} + \gamma^{-2}p'DD'p + p'Bu + u'Bp + u'u = 0, \qquad q(t_f) = 0. \qquad (8.5b)$$

Here $p(\cdot)$ is defined over $[0, t_f]$, since it satisfies a linear differential equation with bounded coefficients, and $q(\cdot)$ is merely a definite integral. Now, a direct calculation using (8.1), (8.4) and (8.5) yields

$$\frac{d}{dt}(x'Sx + p'x + x'p + q) = -\gamma^2|w - \gamma^{-2}D'(Sx + p)|^2 - |x|_Q^2 - |u|^2 + \gamma^2|w|^2.$$

Notice also that because of the final conditions in (8.4) and (8.5),

$$x'(t_f)S(t_f)x(t_f) + p'(t_f)x(t_f) + x'(t_f)p(t_f) + q(t_f) = |x(t_f)|_{Q_f}^2. \qquad (8.6)$$

Therefore, by integrating the previous identity from 0 to t_f, we arrive at, $\forall w \in \mathcal{W}$,

$$J_\gamma(x_0; u, w) = |x_0|_{S(0)}^2 + 2p'(0)x_0 + q(0) - \gamma^2 \int_0^{t_f} |w - \gamma^{-2}D'(Sx + p)|^2 dt.$$

The right-hand side contains "initial terms" which are independent of w, and an integral which is always nonpositive. Hence J_γ is maximized with respect to w by chosing $w = \gamma^{-2}D'(Sx + p)$, which makes the integral equal to zero. This leads to

$$\max_w J_\gamma(x_0; u, w) = |x_0|_{S(0)}^2 + 2p'(0)x_0 + q(0), \qquad (8.7)$$

which is finite. ◇

Remark 8.1. The result of Theorem 8.1 readily generalizes to a payoff J_γ that also includes a linear term in $x(t_f)$ in addition to the quadratic one. In this case, it suffices to change the final condition in (8.5a) to recover (8.6) with the linear term added to the right-hand side. ◇

Proposition 8.2. *For γ sufficiently large, equation (8.4) has a solution over $[0, t_f]$.*

Proof. Replace γ^{-2}, by ϵ in (8.4). For $\epsilon = 0$, this is a Lyapunov equation, and thus it admits a solution S over $[0, t_f]$. The solution of (8.2), however, is a continuous function of ϵ, at zero as elsewhere. Therefore it remains bounded for small enough values of ϵ, or equivalently for large enough values of γ. ◇

As a consequence of the above, the following set is nonempty:

$$\widehat{\Gamma} = \{\widetilde{\gamma} > 0 \mid \forall \gamma \geq \widetilde{\gamma}, \text{ the RDE (8.4) has a solution over } [0, t_f]\} \quad (8.8a)$$

Define $\widehat{\gamma}$ as

$$\widehat{\gamma} = \inf \left\{ \gamma \in \widehat{\Gamma} \right\}. \quad (8.8b)$$

We now prove the following fact:

Proposition 8.3. *For $\gamma = \widehat{\gamma}$, the RDE (8.4) has a finite escape time.*

Proof. For any (γ, t) for which $S_\gamma(t)$ exists, it is a continuous function of γ, and thus so is $\dot{S}_\gamma(t)$ according to (8.4). As a consequence, S_γ is continuous in γ uniformly in t over $[0, t_f]$. As a matter of fact, given an $\epsilon > 0$, if we associate a $\delta(t) \in \mathbb{R}$ with each $t \in [0, t_f]$, such that for $\widetilde{\gamma} \in (\gamma - \delta, \gamma + \delta)$, $|S_{\widetilde{\gamma}}(t) - S_\gamma(t)| < \epsilon/2$, and $\|\dot{S}_{\widetilde{\gamma}}(t) - \dot{S}_\gamma(t)\| < \epsilon/2$, then there exists a neighborhood of t so that for t' in this neighborhood, $|S_{\widetilde{\gamma}}(t') - S_\gamma(t')| < \epsilon$. This defines an open covering of the compact line segment $[0, t_f]$. Extract a finite covering, and pick δ as the minimum of the corresponding δ's. For $\widetilde{\gamma} \in (\gamma - \delta, \gamma + \delta)$, $\forall t \in [0, t_f]$, we have $|S_{\widetilde{\gamma}}(t) - S_\gamma(t)| < \epsilon$. This provides an *a priori* bound for $S_{\widetilde{\gamma}}$, and thus ensures its existence over $[0, t_f]$.

Appendix A: Conjugate Points

Therefore, the set of all γ's for which S_γ is defined over $[0, t_f]$ is open, and $\widehat{\Gamma}$ is its connected component that contains $+\infty$. Hence it is open, and $\widehat{\gamma} \notin \widehat{\Gamma}$. ◇

Remark 8.2. It will follow from the next theorem that $\widehat{\Gamma}$ is in fact the set of all γ's for which S_γ is defined over $[0, t_f]$. ◇

We are now ready to state and prove a weak converse of Theorem 8.1.

Theorem 8.2. *$J_\gamma(x_0; u, w)$ has a finite supremum in w for all x_0 if, and only if, $\gamma > \widehat{\gamma}$.*

Proof. The "if" part has already been proven in Theorem 8.1; therefore we prove here only the "only if" part. Toward this end, suppose that we take $\gamma \leq \widehat{\gamma}$. Let $\{\gamma_k\}_{k \geq 0}$ be a monotonically decreasing sequence in $I\!R^+$ with limit point $\widehat{\gamma}$, and write S_k for S_{γ_k}. We know that for some $t^* \in [0, t_f)$, $\|S_k(t^*)\| \to \infty$ as $k \to \infty$. We now need the following fact:

Lemma 8.1. *There exists a fixed $\bar{x} \in I\!R^n$ such that the sequence $\{|\bar{x}|^2_{S_k}\}_{k>0}$ is unbounded.*

Proof of the Lemma. According to Proposition 8.1, S_k is nonnegative definite for each $k \geq 0$, so that a valid norm, equivalent to any other one, is its trace. Hence $Tr[S_k] \to \infty$. As a consequence, at least one of the diagonal elements is unbounded, since otherwise the trace would be bounded by the (finite) sum of these bounds. Now, picking $\bar{x}$ as the basis vector associated with that diagonal element proves the lemma. Note that at the expense of taking a subsequence of the k's, we may assume that $\|x\|^2_{S_k} \to \infty$. ◇

Now, returning to the proof of Theorem 8.2, we let $\Phi(\cdot,\cdot)$ denote the state transition matrix associated with A, and choose

$$x_0 = \Phi(0,t^*)\bar{x} - \int_0^{t^*} \Phi(0,t)B(t)u(t)dt,$$

so that $w = 0$ for $t \in [0,t^*]$ will yield $x(t^*) = \bar{x}$. For all such w's

$$J_\gamma(x_0;u,w) \geq |x(t_f)|^2_{Q_f} + \int_{t^*}^{t_f} (|x|^2_Q + |u|^2 - \gamma^2|w|^2)dt.$$

In view of (8.7) on the interval $[t^*, t_f]$, this implies $\max_w J_{\gamma_k}(x_0;u,w) \to \infty$. But we also have

$$\forall w \in \mathcal{W}, \quad J_\gamma(x_0;u,w) = J_{\gamma_k}(x_0;u,w) + (\gamma_k^2 - \gamma^2)\|w\|^2 \geq J_{\gamma_k}(x_0;u,w).$$

Consequently, $\sup_w J_\gamma(x_0;u,w) \geq \max_w J_{\gamma_k}(x_0;u,w)$, for all $k \geq 0$. Hence the supremum of J_γ over $w \in \mathcal{W}$ is infinite. ◇

Remark 8.3. The result of Theorem 8.2 is the only one in this appendix that concerns $J_{\widehat{\gamma}}$. ◇

Before we state and prove the next theorem, we need to introduce Caratheodory's *canonic equations* and prove two lemmas. Let $P(\cdot)$ be a symmetric piecewise continuous bounded matrix function on $[0,t_f]$, and let $X(\cdot)$ and $Y(\cdot)$ be two $n \times n$ matrix functions defined by

$$\dot{X} = AX - PY, \qquad X(t_f) = I, \qquad (8.9a)$$

$$\dot{Y} = -QX - A'Y, \qquad Y(t_f) = Q_f. \qquad (8.9b)$$

Notice that X and Y are well defined over $(0,t_f]$, and that $[X'(t), Y'(t)]'$ has a constant rank n. As a matter of fact, it is the transition matrix generating all the solutions of the Hamiltonian system

$$\dot{x} = Ax - P\lambda, \qquad (8.10a)$$

$$\dot{\lambda} = -Qx - A'\lambda, \qquad \lambda(t_f) = Q_f x(t_f) \qquad (8.10b)$$

Appendix A: Conjugate Points

for all possible initial states.

Consider also the associated Riccati differential equation in K:

$$\dot{K} + KA + A'K - KPK + Q = 0, \quad K(t_f) = Q_f. \tag{8.11}$$

Lemma 8.2. *The RDE (8.11) has a bounded solution over $[0, t_f]$ if, and only if, the matrix function $X(\cdot)$ that solves (8.9) is invertible over $[0, t_f]$. Otherwise, the finite escape time t^* of (8.11) is the largest $t^* < T$ such that $X(t^*)$ is singular.*

Proof. By continuity, the matrix X is invertible in a left neighborhood of t_f. Let $K(t) = Y(t)X^{-1}(t)$, and check using (8.9) that this matrix function K is indeed the (unique) solution of (8.11) in that neighborhood. Hence if X is invertible over $[0, t_f]$, K is defined over that interval. Conversely, assume that $K(\cdot)$ is defined over $[t, t_f]$, and let $\Phi_K(\cdot, \cdot)$ be the transition matrix associated with $A - PK$. Let $X(t) = \Phi_K(t, t_f)$, and $Y(t) = K(t)X(t)$, and again check directly that they constitute a solution of (8.9). As a transition matrix is always invertible, this completes the proof of the lemma. ◇

Lemma 8.3. *For any fixed $\xi \in \mathbb{R}^n$, let $x(t) = X(t)\xi$ and $\lambda(t) = Y(t)\xi$ be the solution of (8.10) initialized at $x_0 = X(0)\xi$. Then*

$$|x(t_f)|^2_{Q_f} + \|\lambda\|^2_P + \|x\|^2_Q = \xi'Y'(0)X(0)\xi = \xi'X'(0)Y(0)\xi. \tag{8.12}$$

Proof. Compute

$$\frac{d}{dt}(\lambda'x) = -|x|^2_Q - |\lambda|^2_P$$

and integrate this from 0 to t_f to obtain the first expression. Doing the same calculation with $Y'X$ shows that this matrix remains symmetric for all t. (This also explains why $K = YX^{-1}$ is symmetric.) ◇

We are now in a position to state and prove the following theorem, which holds for every initial state x_0, but says nothing about the case $\gamma = \hat{\gamma}$.

Theorem 8.3. *For any fixed $x_0 \in \mathbb{R}^n$ and $u \in \mathcal{U}$, $J_\gamma(x_0; u, w)$ has a finite supremum in $w \in \mathcal{W}$ if $\gamma > \hat{\gamma}$, and only if $\gamma \geq \hat{\gamma}$.*

Proof. The "if" part is Theorem 8.1. We prove here the "only if" part. Notice that the function $w \mapsto x$ is affine, and therefore $w \mapsto J_\gamma(x_0; u, w)$ is quadratic nonhomogeneous. If its homogeneous part is not concave, it has an infinite supremum. This in turn is characterized by the fact that this homogeneous part, $J_\gamma(0; 0, w)$, is positive for some $w = \bar{w}$. As a matter of fact, if this happens, multiplying $\bar{w}$ by a positive number a will multiply the linear terms in any $J_\gamma(x_0; u, \bar{w})$ by a, and the quadratic terms by a^2, so that the quadratic terms will eventually dominate for large values of a. Therefore henceforth, we investigate the case $x_0 = 0$, $u = 0$.

Applying Lemma 8.2 with $P = -\gamma^{-2}DD'$, we associate with this maximization problem a pair of square matrices, $X_\gamma(t)$, $Y_\gamma(t)$. For $\gamma = \hat{\gamma}$, we know that there is a conjugate point at some $t^* \in [0, t_f)$. Therefore, by Lemma 8.2, $X_{\hat{\gamma}}(t^*)$ is singular. Let $\xi \neq 0$ in $\mathbb{R}^n$ be such that $X_{\hat{\gamma}}(t^*)\xi = 0$. Then, because the matrix $[X', Y']'$ is always of rank n, $Y_{\hat{\gamma}}(t^*)\xi \neq 0$, and remains nonzero in a right neighborhood of t^*. Let $\bar{w} \in \mathcal{W}$ be defined by

$$\bar{w}(t) = \begin{cases} 0 & \text{if } t \in [0, t^*) \\ \gamma^{-2}D'Y_{\hat{\gamma}}\xi & \text{if } t \in [t^*, t_f]. \end{cases}$$

Applying Lemma 8.3 between t^* and t_f, (8.12) yields $J_{\hat{\gamma}}(0; 0, \bar{w}) = 0$. Notice also that $\|\bar{w}\|^2 > 0$. Now, take any $\gamma < \hat{\gamma}$. Again we have

$$J_\gamma(0; 0, \bar{w}) = J_{\hat{\gamma}}(0; 0, \bar{w}) + (\hat{\gamma}^2 - \gamma^2)\|\bar{w}\|^2 = (\hat{\gamma}^2 - \gamma^2)\|\bar{w}\|^2 > 0.$$

Hence $J_\gamma(0; 0, w)$ is unbounded, and so is $J_\gamma(x_0; u, w)$ for any $x_0 \in \mathbb{R}^n$ and $u \in \mathcal{U}$. ◇

Remark 8.4. It can further be shown that at $\gamma = \hat{\gamma}$, the conjugate point is in fact at $t = 0$, and that $J_{\hat{\gamma}}(x_0; u, w)$ has a finite supremum if, and only if, $x_0 \in \text{Im} X_{\hat{\gamma}}(0)$. ◇

Appendix A: Conjugate Points

We now turn to the investigation of [1]

$$\sup_w \min_u J_\gamma(x_0; u, w).$$

To this (max min) optimization problem, we now associate a new Riccati differential equation:

$$\dot{Z} + AZ + A'Z - Z(BB' - \gamma^{-2}DD')Z + Q = 0, \quad Z(t_f) = Q_f. \quad (8.13)$$

A finite escape time for (8.13) is also called a *conjugate point*. A proof identical to the one of Proposition 8.1 yields:

Proposition 8.4. *Whenever it exists, $Z_\gamma(t)$ is nonnegative definite, and it is positive definite if either Q_f or $Q(\cdot)$ is positive definite.* ◇

The counterpart to Theorem 8.1 is the following:

Theorem 8.4. *If Z_γ is defined over $[0, t_f]$, then*

$$\max_w \min_u J_\gamma(x_0; u, w) = x_0' Z_\gamma(0) x_0. \quad (8.14)$$

Proof. The proof relies upon the method of *completion of squares*. Using (8.13) and (8.1),

$$\frac{d}{dt}(x'(t) Z_\gamma(t) x(t))$$

$$= |u + B' Z_\gamma x|^2 - \gamma^2 |w - \gamma^{-2} D' Z_\gamma x|^2 - |x|_Q^2 - |u^2| + \gamma^2 |w|^2$$

so that

$$\forall (u, w) \in \mathcal{U} \times \mathcal{W},$$

$$J_\gamma(x_0; u, w) = x_0' Z_\gamma(0) x_0 + \|u + B' Z_\gamma x\|^2 - \gamma^2 \|w - \gamma^{-2} D' Z_\gamma x\|^2. \quad (8.15)$$

---
[1] Proofs of some of the results to follow are quite similar to their counterparts in the pure maximization problem already discussed, and hence they will be not be included here.

This expression shows that if the "players" are allowed to use closed-loop controls, of the form $u(t) = \mu(t, x(t))$, $v(t) = \nu(t, x(t))$, then the pair

$$\mu^*(t, x) = -B'(t)Z_\gamma(t)x, \quad \nu^*(t, x) = \gamma^{-2}D'(t)Z_\gamma(t)x \tag{8.16}$$

is a saddle point. However we are looking here for open-loop controls. Let in the sequel x^* be the trajectory generated by strategies (8.16), and $u^*(t) := \mu^*(t, x^*(t))$, $w^*(t) := \nu^*(t, x^*(t))$ be the open-loop controls on that trajectory. For any fixed w, we may choose the control $u(t)$ that would be generated by μ^* together with that w. We see that then $J_\gamma \leq x_0' Z_\gamma(0) x_0$. Therefore, *a fortiori*,

$$\min_u J_\gamma(x_0; u, w) \leq x_0' Z_\gamma(0) x_0 ,$$

and since this is true for all $w \in \mathcal{W}$,

$$\sup_w \min_u J_\gamma(x_0; u, w) \leq x_0' Z_0(0) x_0. \tag{8.17}$$

Let us now study the optimization problem:

$$\min_u J_\gamma(x_0; u, w^*) .$$

In view of (8.2), its unique solution is given by $u = -B'\lambda$, where λ is the solution of

$$\dot{x} = Ax - BB'\lambda + \gamma^{-2}DD'Z_\gamma x^*, \quad x(0) = x_0$$

$$\dot{\lambda} = -Qx - A'\lambda, \qquad\qquad \lambda(t_f) = Q_f x(t_f).$$

It is easy to check directly that $x = x^*$, $\lambda = Z_\gamma x^* := \lambda^*$ is a solution of these equations. But then the pair (x^*, λ^*) satisfies the Hamiltonian system of the form (8.10), with $P = BB' - \gamma^{-2}DD'$. Therefore, we can apply Lemma 8.3, to arrive at

$$\min_u J_\gamma(x_0; u, w^*) = x_0' Z_\gamma(0) x_0.$$

This, together with (8.17), proves the claim. ◊

A reasoning identical to the one used in the proof of Proposition 8.2 now leads to

Appendix A: Conjugate Points

Proposition 8.5. *For γ large enough, (8.13) has a solution over $[0, t_f]$.* ◇

As a result, we may introduce the following nonempty set as the counterpart of $\widehat{\Gamma}$:

$$\Gamma := \{\tilde{\gamma} \geq 0 \mid \forall \gamma \geq \tilde{\gamma}, \text{ the RDE (8.13) has a solution over } [0, t_f]\}. \tag{8.18a}$$

Furthermore let

$$\gamma^* = \inf\{\gamma \in \Gamma\} \tag{8.18b}$$

Again a proof identical to the one of Proposition 8.3 yields the following result:

Proposition 8.6. *For $\gamma = \gamma^*$, the RDE (8.13) has a conjugate point $t^* \in [0, t_f)$.* ◇

Now, Theorem 8.2 does not readily carry over to the *sup min* problem, because in the time interval $[0, t^*]$, u might be able to drive the state through an $x(t^*)$ for which the value $x'(t^*)Z_\gamma(t^*)x(t^*)$ does not go unbounded as $\gamma \downarrow \gamma^*$. We turn therefore toward establishing the counterpart of Theorem 8.3.

Theorem 8.5. *For any fixed $x_0 \in \mathbb{R}^n$, the function $\min_u J_\gamma(x_0; u, w)$ has a finite supremum in $w \in \mathcal{W}$ if $\gamma > \gamma^*$, and only if $\gamma \geq \gamma^*$.*

Proof. We first note that using (8.2) an expression for $\min_u J_\gamma$ is given by

$$\min_u J_\gamma(x_0; u, w) = |x(t_f)|^2 + \|x\|_Q^2 + \|\lambda\|_{BB'}^2 - \gamma^2 \|w\|^2. \tag{8.19}$$

The function $w \mapsto (x, \lambda)$, in (8.2b)-(8.2c), is still affine, and hence the above expression still defines a quadratic nonhomogeneous function of w. Therefore as in Theorem 8.3 we study only the homogeneous term, which is given by (8.19) with $x_0 = 0$. Let $X_\gamma(t)$, $Y_\gamma(t)$ be the solutions of (8.9) with $P = BB' - \gamma^{-2}DD'$. (Note that in this case (8.11) is in fact (8.13)). Let t^* be the conjugate point corresponding to γ^*, and as in Theorem 8.3 let $\xi \neq 0$ be such that $X_{\gamma^*}(t^*)\xi = 0$, which ensures that $Y_{\gamma^*}(t^*)\xi \neq 0$. The control

$$\bar{w}(t) = \begin{cases} 0 & \text{for } t \in [0, t) \\ \gamma^{*-2}D'Y_{\gamma^*}(t)\xi & \text{for } t \in [t^*, t_f] \end{cases}$$

will lead to $x(t) = X_{\gamma^*}(t)\xi$, $\lambda(t) = Y_{\gamma^*}(t)\xi$ from t^* on, and therefore by applying Lemma 8.3,

$$\min_u J_{\gamma^*}(0; u, \bar{w}) = 0,$$

although $\|\bar{w}\| > 0$. Therefore, as previously, for any $\gamma < \gamma^*$, $\min_u J_\gamma(0; u, \bar{w}) > 0$, and thus $\sup_w \min_u J_{\gamma^*}(x_0; u, w) = \infty$. ◇

Although Theorem 8.5 seems to be comparable to Theorem 8.3, the situation is in fact significantly more complicated in the present case than in the case of simple maximization. What happens at $\gamma = \gamma^*$ depends on further properties of the conjugate point. See [21] for a more complete analysis. The following example is an extension of the example first presented in [21], to show some of these intricacies.

Example 8.1. Let $n = 1$, $t_f = 2$, and the game be described by

$$\dot{x} = (2 - t)u + tw \quad x(0) = x_0$$

$$J_\gamma(x_0; u, w) = \frac{1}{2}x(2)^2 + \|u\|^2 - \gamma^2\|w\|^2.$$

The associated Riccati differential equation (8.13) is

$$\dot{Z} + (\gamma^2 - 1)Z^2 = 0, \quad Z(2) = \frac{1}{2}.$$

It admits the unique solution

$$Z_\gamma(t) = \left[2(t-1)^2 + \frac{1 - \gamma^{-2}}{3}(8 - t^3)\right]^{-1}$$

Appendix A: Conjugate Points

provided that the inverse exists. For any $\gamma > 1$, $Z_\gamma(t)$ is positive and hence is defined on $[0, 2]$. However, for $\gamma = \gamma^* = 1$, a conjugate point appears at $t^* = 1$, i.e., in the interior of $[0, 2]$. Moreover, for $\gamma = 1$, the feedback control

$$u(t) = \mu(t, x(t)) = -\frac{2-t}{2(t-1)^2} x(t),$$

although defined by a gain that diverges at $t = 1$, will produce a bounded control trajectory for every $w \in \mathcal{W}$, so that the quantity

$$\sup_w \inf_u J_{\gamma^*}(x_0; u, w)$$

is bounded for every $x_0 \in \mathbb{R}$. ◊

Chapter 9

Appendix B: Danskin's Theorem

In this appendix we state and prove a theorem due to Danskin, which was used in Chapter 5, in the proof of Theorem 5.1. As far as we know, the original version was first presented in [33]. Since then, many more general forms have been derived. More discussion on this can be found in [24]. To follow the arguments used in this chapter, some background in real analysis, for example at the level of [72], is required.

We first introduce some preliminaries:

Let $I = [0, t_f] \subset \mathbb{R}$, and Ω be a Hilbert space, with inner product $(\cdot, \cdot)$, and norm $\|\cdot\|$. Let $G : I \times \Omega \to \mathbb{R} : (t, \omega) \mapsto G(t, \omega)$ be given, and define

$$W(t) := \sup_{\omega \in \Omega} G(t, \omega).$$

The hypotheses on G are as follows.

Hypothesis H1. $\forall t \in I$, $\omega \mapsto G(t, \omega)$ is concave, upper semicontinuous, and $G(t, \omega) \to -\infty$ as $\omega \to \infty$. ◇

As a consequence of hypothesis H1,

$$\forall t \in I, \quad \exists \, \widehat{\omega}(t) \in \Omega : \quad G(t, \widehat{\omega}(t)) = W(t).$$

Hypothesis H2.

$$\forall t \in I, \quad \exists \, \nu > 0 : \quad \forall \omega \in \Omega, \quad W(t) - G(t, \omega) \geq \nu \|\omega - \widehat{\omega}(t)\|^2. \quad \diamond$$

Hypothesis H3. $\forall \omega \in \Omega$, $\forall t \in (0, t_f)$, there exists a partial derivative

$$\frac{\partial}{\partial t} G(t, \omega) =: \dot{G}(t, \omega)$$

$(t, \omega) \mapsto \dot{G}(t, \omega)$ is continuous, $\dot{G}$ being bounded over any bounded subset of Ω. (For $t = 0$ or t_f we may assume one-sided derivatives). ◊

As a consequence of hypothesis H3, $t \mapsto G(t, \omega)$ is continuous, uniformly in ω over any bounded subset of $I \times \Omega$.

Remark 9.1. For a quadratic form

$$G(t, \omega) = -(\omega, A(t)\omega) + 2(b(t), \omega) + c(t),$$

the hypothesis H2 is a consequence of H1. As a matter of fact, concavity requires that A be positive definite, and the hypothesis

$$G(t, \omega) \to -\infty \text{ as } \omega \to \infty$$

requires that A be coercive (or elliptic), i.e.,

$$\inf_{\|\omega\|=1} (\omega, A\omega) = \nu > 0. \tag{9.1}$$

If this infimum were zero, then letting $\{\omega_n\}$ be a sequence with $\|\omega_n\| = 1$ and $\lim (\omega_n, A\omega_n) = 0$, we could choose $\tilde{\omega}_n = \epsilon_n (\omega_n, A\omega_n)^{-\frac{1}{2}} \omega_n$, with $\epsilon_n = \pm 1$ such that $(b, \tilde{\omega}_n) \geq 0$. Then $\tilde{\omega}_n$ would diverge to infinity, and yet $G(t, \omega) \geq c(t) - 1$.

But then, if A is coercive, it is invertible, $\hat{\omega}(t) = A^{-1}(t) b(t)$, and

$$W(t) = \left(b(t), A^{-1}(t) b(t)\right) + c(t), \tag{9.2a}$$

$$G(t, \omega) = -(\omega - \hat{\omega}(t), A(t)[\omega - \hat{\omega}(t)]) + (b(t), A^{-1}(t) b(t)) + c(t) \tag{9.2b}$$

(9.1) and (9.2) together prove that H2 is satisfied. ◊

We now state and prove the main theorem of this appendix.

Theorem 9.1. *Under the hypotheses H1-H3, W has a derivative given by*

$$\frac{dW(t)}{dt} = \dot{G}(t, \hat{\omega}(t))$$

Proof. We shall look separately at the right derivative and the left derivative, so that the proof will hold at $t = 0$ and $t = t_f$. For convenience, define $\hat{\omega} = \hat{\omega}(t)$, and

$$\Delta(h) = \frac{1}{h}[W(t+h) - W(t)].$$

Let $\{h_n\}$ be a sequence of positive numbers decreasing to zero. From the definition of W, we have

$$\Delta(h_n) = \frac{1}{h_n}[W(t+h_n) - G(t,\hat{\omega})] \geq \frac{1}{h_n}[G(t+h_n,\hat{\omega}) - G(t,\hat{\omega})]$$

so that

$$\liminf_{h_n \downarrow 0} \Delta(h_n) \geq \dot{G}(t,\hat{\omega}). \tag{9.3}$$

Now let $\omega_n := \hat{\omega}(t+h_n)$. By H1, the ω_n's stay in a bounded set. Hence the sequence has at least one weak accumulation point $\bar{\omega}$. Take a subsequence, again noted $\{\omega_n\}$, such that $\omega_n \to \bar{\omega}$ (ω_n converges weakly to $\bar{\omega}$). We then have

$$G(t+h_n, \omega_n) - G(t,\bar{\omega}) = G(t+h_n,\omega_n) - G(t,\omega_n) + G(t,\omega_n) - G(t,\bar{\omega}).$$

The uniform continuity of G with respect to t yields

$$G(t+h_n,\omega_n) - G(t,\omega_n) \to 0 \quad \text{as } n \to \infty.$$

G, being a concave function of ω and upper semicontinuous, is also upper semicontinuous in the weak topology. Thus

$$\limsup_{n \to \infty} [G(t,\omega_n) - G(t,\bar{\omega})] \leq 0.$$

From the above three relations, we conclude that

$$\limsup_{n \to \infty} G(t+h_n,\omega_n) \leq G(t,\bar{\omega}). \tag{9.4}$$

Now, it follows from (9.3) that there exists a real number a such that $\Delta(h_n) \geq a$, or equivalently (recall that $h_n > 0$)

$$G(t+h_n,\omega_n) \geq G(t,\hat{\omega}) + ah_n.$$

Therefore
$$\liminf_{h_n \downarrow 0} G(t+h_n, \omega_n) \geq G(t, \widehat{\omega}). \tag{9.5}$$
It follows from (9.4) and (9.5) that
$$G(t, \bar{\omega}) \geq G(t, \widehat{\omega}).$$
This implies that $\bar{\omega} = \widehat{\omega}$, $G(t, \bar{\omega}) = G(t, \widehat{\omega})$, and
$$G(t+h_n, \omega_n) \to G(t, \widehat{\omega}).$$
Using H2, together with the uniform continuity of G with respect to t, we get
$$\nu \|\omega_n - \widehat{\omega}\|^2 \leq G(t, \widehat{\omega}) - G(t, \omega_n)$$
$$= G(t, \widehat{\omega}) - G(t+h_n, \omega_n) + G(t+h_n, \omega_n) - G(t, \omega_n) \to 0,$$
which in turn implies that $\omega_n \to \widehat{\omega}$ in the strong topology of Ω.

Finally, we have
$$\Delta(h_n) \leq \frac{1}{h_n}[G(t+h_n, \omega_n) - G(t, \omega_n)].$$
However, G is differentiable in t for all t in I. Thus, by the mean-value theorem, there exists a $\theta_n \in [0, 1]$ such that
$$\frac{1}{h_n}[G(t+h_n, \omega_n) - G(t, \omega_n)] = \dot{G}(t+\theta_n h_n, \omega_n).$$
By the continuity of $\dot{G}$, it follows that
$$\liminf_{h_n \downarrow 0} \Delta(h_n) \leq \dot{G}(t, \widehat{\omega}). \tag{9.6}$$
The inequalities (9.3) and (9.6) together imply that
$$\lim_{h_n \downarrow 0} \frac{1}{h_n}[W(t+h_n) - W(t)] = \dot{G}(t, \widehat{\omega}).$$
Since the derivation was carried out for $h_n > 0$, this shows that W has a right derivative equal to $\dot{G}(t, \widehat{\omega})$.

The proof for the left derivative is entirely similar, with $h_n < 0$. The first argument of the above proof now gives, instead of (9.3):
$$\limsup_{h_n \uparrow 0} \Delta(h_n) \leq \dot{G}(t, \widehat{\omega}). \tag{9.7}$$

The proof leading to (9.4) does not depend on the sign of h_n. The inequality (9.7) yields

$$\Delta(h_n) \leq a$$

and h_n being now negative, this gives again (9.5), and hence the strong convergence of ω_n to $\widehat{\omega}$. Then, the last argument of the earlier proof leads to

$$\liminf_{h_n \uparrow 0} \Delta(h_n) \geq \dot{G}(t,\widehat{\omega}) \qquad (9.8)$$

and this, together with (9.7), completes the proof. ◇

Chapter 10

References

[1] B. D. O. Anderson and J. B. Moore. *Optimal Filtering.* Prentice-Hall, Englewood Cliffs, NJ, 1979.

[2] B. D. O. Anderson and J. B. Moore. *Optimal Control: Linear Quadratic Methods.* Prentice-Hall, Englewood Cliffs, NJ, 1990.

[3] M. Athans, editor. *Special Issue on Linear-Quadratic-Gaussian Control,* volume 16(6). IEEE Trans. Automat. Control, 1971.

[4] T. Başar. A general theory for Stackelberg games with partial state information. *Large Scale Systems,* 3(1):47–56, 1982.

[5] T. Başar. Disturbance attenuation in LTI plants with finite horizon: Optimality of nonlinear controllers. *Systems and Control Letters,* 13(3):183–191, September 1989.

[6] T. Başar. A dynamic games approach to controller design: Disturbance rejection in discrete time. In *Proceedings of the 29th IEEE Conference Decision and Control,* pages 407–414, December 13-15, 1989. Tampa, FL.

[7] T. Başar. Time consistency and robustness of equilibria in noncooperative dynamic games. In F. Van der Ploeg and A. de Zeeuw, editors, *Dynamic Policy Games in Economics,* pages 9–54. North Holland, 1989.

[8] T. Başar. Game theory and H^∞-optimal control: The continuous-time case. In *Proceedings of the 4th International Conference on Differential Games and Applications.* Springer-Verlag, August 1990. Helsinki, Finland.

[9] T. Başar. Game theory and H^∞-optimal control: The discrete-time case. In *Proceedings of the 1990 International Conference on New Trends in Communication, Control and Signal Processing,* pages 669–686, Ankara, Turkey, July 1990. Elsevier.

[10] T. Başar. H^∞-Optimal Control: A Dynamic Game Approach. University of Illinois, 1990.

[11] T. Başar. Minimax disturbance attenuation in LTV plants in discrete-time. In *Prooceedings of the 1990 Automatic Control Conference*, San Diego, May 1990.

[12] T. Başar. Optimum performance levels for minimax filters, predictors and smoothers. Preprint, May 1990.

[13] T. Başar. Generalized Riccati equations in dynamic games. In S. Bittanti, A. Laub, and J. C. Willems, editors, *The Riccati Equation*. Springer-Verlag, March 1991.

[14] T. Başar. Optimum H^∞ designs under sampled state measurements. To appear in *Systems and Control Letters*, 1991.

[15] T. Başar and P. R. Kumar. On worst case design strategies. *Computers and Mathematics with Applications: Special Issue on Pursuit – Evasion Differential Games*, 13(1-3):239–245, 1987.

[16] T. Başar and M. Mintz. Minimax terminal state estimation for linear plants with unknown forcing functions. *International Journal of Control*, 16(1):49–70, August 1972.

[17] T. Başar and G. J. Olsder. *Dynamic Noncooperative Game Theory*. Academic Press, London/New York, 1982.

[18] M. D. Banker. *Observability and Controllability of Two-Player Discrete Systems and Quadratic Control and Game Problems*. PhD thesis, Stanford University, 1971.

[19] A. Bensoussan. Saddle points of convex concave functionals. In H. W. Kuhn and G. P. Szegö, editors, *Differential Games and Related Topics*, pages 177–200. North-Holland, Amsterdam, 1971.

[20] P. Bernhard. Sur la commandabilitè des systèmes linèaires discrete à deux joueurs. *Rairo*, J3:53–58, 1972.

[21] P. Bernhard. Linear-quadratic two-person zero-sum differential games: necessary and sufficient conditions. *Journal of Optimization Theory & Applications*, 27:51–69, 1979.

[22] P. Bernhard. Exact controllability of perturbed continuous-time linear systems. *IEEE Transactions on Automatic Control*, AC-25(1):89–96, 1980.

[23] P. Bernhard. A certainty equivalence principle and its applications to continuous-time sampled data, and discrete time H^∞ optimal control. INRIA Report #1347, August 1990.

[24] P. Bernhard. Variations sur un thème de Danskin avec une coda sur un thème de Von Neumann. INRIA Report #1238, March 1990.

[25] P. Bernhard. Application of the min-max certainty equivalence principle to sampled data output feedback H^∞ optimal control. To appear in *Systems and Control Letters*, 1991.

[26] P. Bernhard and G. Bellec. On the evaluation of worst case design with an application to the quadratic synthesis technique. In *Proceedings of the 3rd IFAC Symposium on Sensitivity, Adaptivity, and Optimality*, Ischia, Italy, 1973.

[27] D. P. Bertsekas and I. B. Rhodes. On the minimax reachability of target sets and target tubes. *Automatica*, 7:233–247, 1971.

[28] D. P. Bertsekas and I. B. Rhodes. Recursive state estimation for a set-membership description of uncertainty. *IEEE Transactions on Automatic Control*, AC-16:117–128, April 1971.

[29] D. P. Bertsekas and I. B. Rhodes. Sufficiently informative functions and the minimax feedback control of uncertain dynamic systems. *IEEE Transactions on Automatic Control*, AC-18(2):117–123, April 1973.

[30] C. C. Caratheodory. *Calculus of Variations and Partial Differential Equations of the First order*. Holden-Day, New York, NY, 1967. Original German edition published by Teubner, Berlin, 1935.

[31] B. C. Chang and J. B. Pearson. Optimal disturbance reduction in linear multivariable systems. *IEEE Transactions on Automatic Control*, AC-29:880–887, October 1984.

[32] J. B. Cruz, Jr. *Feedback Systems*. McGraw Hill, New York, NY, 1971.

[33] J. M. Danskin. *The Theory of Max Min*. Springer, Berlin, 1967.

[34] G. Didinsky and T. Başar. Design of minimax controllers for linear systems with nonzero initial conditions and under specified information structures. In *Proceedings of the 29th IEEE Conference on Decision and Control*, pages 2413–2418, Honolulu, HI, December 1990.

[35] P. Dorato, editor. *Robust Control*. IEEE Press, New York, NY, 1987.

[36] P. Dorato and R. F. Drenick. Optimality, insensitivity, and game theory. In L. Radanović, editor, *Sensitivity Methods in Control Theory*, pages 78–102. Pergamon Press, New York, NY, 1966.

[37] J. Doyle, K. Glover, P. Khargonekar, and B. Francis. State-space solutions to standard H_2 and H_∞ control problems. *IEEE Transactions on Automatic Control*, AC-34(8):831–847, 1989.

[38] T. S. Ferguson. *Mathematical Statistics, A Decision Theoretic Approach*. Academic Press, New York, 1967.

[39] B. A. Francis. *A Course in H_∞ Control Theory*, volume 88 of *Lecture Notes in Control and Information Sciences*. Springer-Verlag, New York, 1987.

[40] B. A. Francis and J. C. Doyle. Linear control theory with an H_∞ optimality criterion. *SIAM J. Control Optimization*, 25:815–844, 1987.

[41] B. A. Francis, J. W. Helton, and G. Zames. H_∞-optimal feedback controllers for linear multivariable systems. *IEEE Trans. Automat. Control*, 29:888–9000, 1984.

[42] P. M. Frank. *Introduction to System Sensitivity Theory*. Academic Press, New York, 1978.

[43] K. Glover and J. C. Doyle. State-space formulae for all stabilizing controllers that satisfy an H^∞-norm bound and relations to risk sensitivity. *Systems and Control Letters*, 11:167–172, 1988.

[44] K. Glover, D. J. N. Limebeer, J. C. Doyle, E. Kasenally, and M. G. Safonov. A characterization of all solutions to the four-block general distance problem. *SIAM Journal on Control*, 1991.

[45] J. W. Helton and J. A. Ball. H^∞ control for nonlinear plants: Connectors with differential games. In *Proceedings of the 29th IEEE Conference Decision and Control*, pages 956–962, December 13-15, 1989. Tampa, FL.

[46] M. Heymann, M. Pachter, and R. J. Stern. Max-min control problems: A system theoretic approach. *IEEE Transactions on Automatic Control*, AC-21(4):455–463, 1976.

[47] M. Heymann, M. Pachter, and R. J. Stern. Weak and strong max-min controllability. *IEEE Trans. Automat. Control*, 21(4):612–613, 1976.

[48] P. J. Huber. *Robust Statistics*. Wiley, New York, NY, 1981.

[49] R. Isaacs. *Differential Games*. Kruger Publishing Company, Huntington, NY, 1975. First edition: Wiley, NY 1965.

[50] P. P. Khargonekar. State-space H_∞ control theory and the LQG control problem. In A. C. Antoulas, editor, *Mathematical System Theory: The Influence of R. E. Kalman*. Springer Verlag, 1991.

[51] P. P. Khargonekar, K. Nagpal, and K. Poolla. H^∞-optimal control with transients. *SIAM J. of Control and Optimization*, 1991.

[52] P. P. Khargonekar and K. M. Nagpal. Filtering and smoothing in an H_∞ setting. In *Proceedings of the 29th IEEE Conference on Decision and Control*, pages 415–420, 1989. Tampa, FL.

[53] P. P. Khargonekar, I. R. Petersen, and M. A. Rotea. H_∞-optimal control with state-feedback. *IEEE Transactions on Automatic Control*, AC-33(8):786–788, 1988.

[54] P. R. Kumar and P. P. Varaiya. *Stochastic Systems: Estimation, Identification and Adaptive Control*. Prentice-Hall, Englewood Cliffs, NJ, 1986.

[55] H. Kwakernaak. Progress in the polynomial solution of the standard H_∞ optimal control problem. In V. Utkin and O. Jaaksoo, editors, *Proceedings of the 11th IFAC World Congress*, volume 5, pages 122–134, August 13-17 1990. Tallinn, Estonia, USSR.

[56] H. Kwakernaak and R. Sivan. *Linear Optimal Control Systems*. Wiley-Interscience, New York, 1972.

[57] I. D. Landau. *Adaptive Control: The Model Reference Approach*. Marcel Dekker, New York, 1979.

[58] G. Leitmann. Guaranteed asymptotic stability for some linear systems with bounded uncertainties. *ASME Journal Dynamic Systems, Measurement, and Control*, 101(3), 1979.

[59] D. J. N. Limebeer, B. D. O. Anderson, P. Khargonekar, and M. Green. A game theoretic approach to H_∞ control for time varying systems. In *Proceedings of the International Symposium on the Mathematical Theory of Networks and Systems*, Amsterdam, Netherlands, 1989.

[60] D. J. N. Limebeer, M. Green, and D. Walker. Discrete time H_∞ control. In *Proceedings of the 29th IEEE Conference on Decision and Control*, pages 392–396, Tampa, FL, 1989.

[61] D. J. N. Limebeer and G. D. Halikias. A controller degree bound for H^∞-optimal control problems of the second kind. *SIAM Journal on Control and Optimization*, 26:646–667, 1988.

[62] D. J. N. Limebeer, E. Kasenally, I. Jaimouka, and M. G. Safonov. All solutions to the four-block general distance problem. In *Proceedings of the 27th IEEE Conference on Decision and Control*, pages 875–880, Austin, Texas, 1988.

[63] D. P. Looze, H. V. Poor, K. S. Vastola, and J. C. Darragh. Minimax control of linear stochastic systems with noise uncertainty. *IEEE Transactions on Automatic Control*, AC-28(9):882–887, September 1983.

[64] E. F. Mageirou. Values and strategies for infinite duration linear quadratic games. *IEEE Transactions on Automatic Control*, AC-21(4):547–550, August 1976.

[65] E. F. Mageirou and Y. C. Ho. Decentralized stabilization via game theoretic methods. *Automatica*, 13:393–399, 1977.

[66] C. J. Martin and M. Mintz. Robust filtering and prediction for linear systems with uncertain dynamics: A game-theoretic approach. *IEEE Trans. Automat. Control*, 28(9):888–896, 1983.

[67] P. H. McDowell and T. Başar. Robust controller design for linear stochastic systems with uncertain parameters. In *Proceedings of the 1986 American Control Conference*, pages 39–44, Seattle, WA, June 1986.

[68] M. Mintz and T. Başar. On tracking linear plants under uncertainty. In *Proceedings of the 3rd IFAC Conference on Sensitivity, Adaptivity and Optimality*, Ischia, Italy, June 1973.

[69] I. R. Petersen. Disturbance attenuation and H_∞ optimization: A design method based on the algebraic Riccati equation. *IEEE Transactions on Automatic Control*, AC-32(5):427–429, May 1987.

[70] B. D. Pierce and D. D. Sworder. Bayes and minimax controllers for a linear system with stochastic jump parameters. *IEEE Trans. Automat. Control*, 16(4):300–306, 1971.

[71] I. Rhee and J. L. Speyer. A game theoretic controller and its relationship to H^∞ and linear-exponential Gaussian synthesis. In *Proceedings of the 29th IEEE Conference Decision and Control*, pages 909–915, December 13-15, 1989. Tampa, FL.

[72] H. L. Royden. *Real Analysis*. MacMillan, 1968.

[73] M. G. Safonov, E. A. Jonckhere, M. Verma, and D. J. N. Limebeer. Synthesis of positive real multivariable feedback systems. *International J. Control*, 45(3):817–842, 1987.

[74] M. G. Safonov, D. J. N. Limebeer, and R. Y. Chiang. Simplifying the H_∞ theory via loop shifting, matrix pencil, and descriptor concepts. *International Journal of Control*, 50:2467–2488, 1989.

[75] D. M. Salmon. Minimax controller design. *IEEE Trans. Automat. Control*, 13(4):369–376, 1968.

[76] L. J. Savage. *The Foundation of Statistics*. Wiley, New York, NY, 1954.

[77] A. Stoorvogel. The discrete time H_∞ control problem: the full information case problem. To appear in *SIAM Journal on Control*, 1991.

[78] A. A. Stoorvogel. *The H_∞ Control Problem: A State Space Approach*. PhD thesis, University of Eindhoven, The Netherlands, October 1990.

[79] D. Sworder. *Optimal Adaptive Control Systems*. Academic Press, New York, 1966.

[80] G. Tadmor. H_∞ in the time domain: the standard four block problem. To appear in *Math. Contr. Sign. & Syst.*, 1991.

[81] K. Uchida and M. Fujita. On the central controller: Characterizations via differential games and LEQG control problems. *Systems and Control Letters*, 13(1):9–13, 1989.

[82] K. Uchida and M. Fujita. Finite horizon H^∞ control problems with terminal penalties. Preprint, April 1990.

[83] R. J. Veillette, J. V. Medanić, and W. R. Perkins. Robust control of uncertain systems by decentralized control. In V. Utkin and O. Jaaksoo, editors, *Proceedings of the 11th IFAC World Congress*, volume 5, pages 116–121, August 13-17 1990. Tallinn, Estonia, USSR.

[84] S. Verdú and H. V. Poor. Minimax linear observers and regulators for stochastic systems with uncertain second-order statistics. *IEEE Trans. Automat. Control*, 29(6):499–511, 1984.

[85] S. Verdú and H. V. Poor. Abstract dynamic programming models under commutative conditions. *SIAM J. Control and Optimization*, 25(4):990–1006, July 1987.

[86] S. Weiland. *Theory of Approximation and Disturbance Attenuation for Linear Systems*. PhD thesis, University of Gronigen, The Netherlands, January 1991.

[87] J. C. Willems. *The Analysis of Feedback Systems*. MIT Press, Cambridge, MA, 1971.

[88] J. C. Willems. Least squares stationary optimal control and the algebraic Riccati equations. *IEEE Transactions on Automatic Control*, AC-16(6):621–634, December 1971.

[89] D. J. Wilson and G. Leitmann. Minimax control of systems with uncertain state measurements. *Applied Mathematics & Optimization*, 2(4):315–336, 1976.

[90] H. S. Witsenhausen. Minimax control of uncertain systems. Report ESL-R-269, MIT, Cambridge, MA, May 1966.

[91] H. S. Witsenhausen. A minimax control problem for sampled linear systems. *IEEE Transactions for Automatic Control*, AC-13:5–21, February 1968.

[92] I. Yaesh and U. Shaked. Minimum H^∞-norm regulation of linear discrete-time systems and its relation to linear quadratic difference games. In *Proceedings of the 29th IEEE Conference Decision and Control*, pages 942–947, December 13-15, 1989. Tampa, FL.

[93] G. Zames. Feedback and optimal sensitivity: Model reference transformation, multiplicative seminorms and approximate inverses. *IEEE Transactions on Automatic Control*, AC-26:301–320, 1981.

Chapter 11

List of Corollaries, Definitions, Lemmas, Propositions, Remarks and Theorems

Chapter 2

Definitions		Propertites		Theorems	
No. 2.1.	p. 19	No. 2.1	p. 16	No. 2.1	p. 15
2.2.	21			2.2	16
2.3.	22			2.3	16
2.4.	23			2.4	20
2.5.	27			2.5	21
				2.6	26

Chapter 3

Lemmas		Propositions		Remarks	
No. 3.1.	p. 30	No. 3.1.	p. 51	No. 3.1.	p. 35
3.2.	45			3.2.	36
3.3.	55			3.3.	47
3.4.	57			3.4.	53
3.5.	58			3.5.	64
3.6.	59			3.6.	64

Theorems	
No. 3.1.	p. 31
3.2.	32
3.3.	37
3.4.	41
3.5.	46
3.6.	52
3.7.	60
3.8.	63
3.9.	70

Chapter 4

Lemmas		*Remarks*		*Theorems*	
No. 4.1.	p. 74	No. 4.1.	p. 83	No. 4.1.	p. 75
		4.2.	83	4.2.	78
		4.3.	88	4.3.	81
		4.4.	105	4.4.	86
		4.5.	106	4.5.	87
		4.6.	107	4.6.	90
				4.7.	93
				4.8.	96
				4.9.	100
				4.10.	103
				4.11.	104
				4.12.	105
				4.13.	106
				4.14.	112

Chapter 5

Corollaries		*Lemmas*		*Propositions*	
No. 5.1.	p. 120	No. 5.1.	p. 123	No. 5.1.	p. 126
5.2.	122	5.2.	135	5.2.	130
5.3.	143			5.3.	132

Remarks		*Theorems*	
No. 5.1.	p. 118	No. 5.1.	p. 119
5.2.	123	5.2.	122
5.3.	126	5.3.	127
5.4.	130	5.4.	137
5.5.	136	5.5.	142
5.6.	136	5.6.	146
		5.7.	148

Chapter 6

Corollaries
No. 6.1.   p. 157

Remarks
No. 6.1.   p. 158
6.2.   161
6.3.   165

Lemmas
No. 6.1.   p. 156
6.2.   158
6.3.   163
6.4.   164

Theorems
No. 6.1.   p. 155
6.2.   157
6.3.   161
6.4.   172
6.5.   173
6.6.   175
6.7.   177

Propositions
No. 6.1.   p. 166

Chapter 7

Corollaries
No. 7.1.   186

Theorems
No. 7.1.   p. 185
7.2.   188
7.3.   190
7.4.   191
7.5.   192

Lemmas
No. 7.1.   p. 183

Remarks
No. 7.1.   p. 186
7.2.   186

Chapter 8

Definitions
No. 8.1.   p. 196

Remarks
No. 8.1.   p. 197
8.2.   199
8.3.   200
8.4.   202

Lemmas
No. 8.1.   p. 199
8.2.   201
8.3.   201

Theorems
No. 8.1.   p. 196
8.2.   199
8.3.   202
8.4.   203
8.5.   205

Propositions
No. 8.1.   p. 196
8.2.   198
8.3.   198
8.4.   203
8.5.   205
8.6.   205

Chapter 9

Remarks *Theorems*
No. 9.1.   p. 209   No. 9.1.   p. 209

Systems & Control: Foundations & Applications

Series Editor

Christopher I. Byrnes
Department of Systems Science and Mathematics
Washington University
Campus P.O. 1040
One Brookings Drive
St. Louis, MO 63130-4899
U.S.A.

Systems & Control: Foundations & Applications publishes research monographs and advanced graduate texts dealing with areas of current research in all areas of systems and control theory and its applications to a wide variety of scientific disciplines.

We encourage preparation of manuscripts in such forms as LaTex or AMS TeX for delivery in camera-ready copy which leads to rapid publication, or in electronic form for interfacing with laser printers or typesetters.

Proposals should be sent directly to the editor or to: Birkhäuser Boston, 675 Massachusetts Avenue, Suite 601, Cambridge, MA 02139, U.S.A.

Estimation Techniques for Distributed Parameter Systems
H.T. Banks and K. Kunisch

Set-Valued Analysis
Jean-Pierre Aubin and Hélène Frankowska

Weak Convergence Methods and Singularly Perturbed Stochastic Control and Filtering Problems
Harold J. Kushner

Methods of Algebraic Geometry in Control Theory: Part I
Scalar Linear Systems and Affine Algebraic Geometry
Peter Falb

H^∞-Optimal Control and Related Minimax Design Problems
A Dynamic Approach
Tamer Başar and Pierre Bernhard